COMPUTATIONAL AND EXPERIMENTAL STUDIES OF ACOUSTIC WAVES

Edited by **Mahmut Reyhanoglu**

Computational and Experimental Studies of Acoustic Waves
http://dx.doi.org/10.5772/65135
Edited by Mahmut Reyhanoglu

Contributors

Jingtao Du, Yang Liu, Long Liu, Zine El Abiddine Fellah, Mohamed Fellah, Claude Depollier, Erick Ogam, Farid G Mitri, Olga Khachay, Andrey Khachay, Yanfei Wang, Shuhong Gong, Yu Liu, Muyu Hou, Lixin Guo, Aamir Shahzad, Miguel Bello-Jiménez, Gustavo Ramírez-Meléndez, Erika Hernández-Escobar, Andrés Camarillo-Avilés, Olivier Jean Michel Pottiez, Christian Cuadrado-Laborde, Antonio Diez, José L. Cruz, Miguel V. Andrés, R. López-Estopier

Published by InTech
Janeza Trdine 9, 51000 Rijeka, Croatia

Publishing Process Manager Marijana Francetic
Technical Editor SPi Global
Cover InTech Design team

First published December, 2017
Printed in Croatia
Legal deposit, Croatia: National and University Library in Zagreb

Additional hard copies can be obtained from orders@intechopen.com

Computational and Experimental Studies of Acoustic Waves, Edited by Mahmut Reyhanoglu
p. cm.
Print ISBN 978-953-51-3715-3
Online ISBN 978-953-51-3716-0

For EU product safety concerns:
IN TECH d.o.o., Prolaz Marije Krucifikse Kozulić 3, 51000 Rijeka, Croatia,
info@intechopen.com or visit our website at intechopen.com.

Meet the editor

Mahmut Reyhanoglu is presently the Glaxo Wellcome Distinguished Professor of Engineering at the University of North Carolina at Asheville, North Carolina, USA. His extensive research makes use of advanced mathematical techniques and models that arise from fundamental physical principles. His major research interests are in the areas of nonlinear dynamical systems and control theory, with particular emphasis on applications to mechanical and aerospace systems. He has authored/coauthored several book chapters and over 120 peer-reviewed journal/proceedings papers. He served on the IEEE Transactions on Automatic Control Editorial Board and on the IEEE Control Systems Society Conference Editorial Board as an associate editor. He also served as International Program Committee member for several conferences and as a member of AIAA Guidance, Navigation, and Control Technical Committee.

Contents

Preface

This book presents recent studies of acoustic wave propagation through different media including the atmosphere, Earth's subsurface, complex dusty plasmas, porous materials, and flexible structures. Mathematical models of the underlying physical phenomena are introduced and studied in detail. With its seven chapters, the book brings together important contributions from renowned international researchers to provide an excellent survey of recent computational and experimental studies of acoustic waves. The first section consists of four chapters that focus on computational studies, while the next section is composed of three chapters that center on experimental studies.

This book is divided into two sections that are focused on the recent studies of acoustic waves. The first section consists of four chapters that focus on applications of computational techniques. Chapter 1 emphasizes application of finite difference (FD) methods to the acoustic wave equation, considering both the first-order velocity-stress acoustic equation and second-order wave equation. To mitigate the effects of grid dispersion, the chapter introduces a new staggered grid FD scheme that improves computational efficiency while preserving high accuracy. The effectiveness of this method is demonstrated via numerical simulations of seismic waves. Chapter 2 is devoted to acoustic analysis of a rectangular acoustic cavity. A unified structural-acoustic coupling analysis framework is introduced for the cavity and its coupled panel structure. The Fourier series with supplementary terms are constructed as the admissible functions, which are smoothed in the whole solution domain including the elastic structural and/or impedance acoustic boundary and coupling interface. All the unknown coefficients and higher-order variables are determined via a Rayleigh-Ritz procedure and differential operation. Numerical examples are then presented to demonstrate the validity and effectiveness of the proposed model. Chapter 3 evaluates wave properties of strongly coupled complex dusty (SCCD) plasmas using a new equilibrium molecular dynamics (EMD) simulation technique. It presents several interesting simulation results, which show that the fluctuation of waves increases with increasing Coulomb coupling parameter and decreases with increasing screening strength. The new results obtained through the EMD method introduced in the chapter for complex dusty plasma are discussed and compared with earlier simulation results based on other numerical methods. It is demonstrated that the proposed model constitutes a highly feasible tool for estimating the behavior of waves in strongly coupled complex dusty plasmas over a range of parameters. Chapter 4 studies the problem of modeling a seismic field acoustic approximation in a layered medium with inclusions of a hierarchical structure, where the inclusion density of each rank differs from the density of the enclosing medium and the elastic parameters coincide with the elastic parameters of the enclosing layer. The chapter also considers the case when the inclusion density of each rank coincides with the density of the host medium and the elastic parameters of the inclusion of each rank differ from the elastic parameters of the enclosing layer.An iterative process of solving the theoretical inverse problem for the case of determining configurations of 2D hierarchical inclusions is developed.

The second section of the book is composed of three chapters that center on experimental studies.

Chapter 5 presents the design and implementation of in-fiber acousto-optic (AO) devices based on acoustic flexural waves. The AO interaction is demonstrated to be an efficient mechanism for the development of AO tunable filters and modulators. The implementation of tapered optical fibers is proposed to shape the spectral response of in-fiber AO devices. Experimental results demonstrate that the geometry of the tapered fiber can be regarded as an extra degree of freedom for the design of AO tunable attenuation filters. It is shown that such filters operate as an amplitude modulator when acoustic reflection is induced. As a particular case, an in-fiber AO modulator composed of a double-ended tapered fiber is reported. Chapter 6 explores wave propagation in porous materials. Acoustic propagation in porous media involves a large number of physical parameters when the structure is elastic. This number is reduced when the structure is rigid because the mechanical part does not intervene with the wave propagation. The study of high and low frequencies separately solves the inverse problem and characterizes the porous materials in the domain of influence of the physical parameters. The proposed methods are simple and effective and allow an acoustic characterization of porous materials using transmitted or reflected experimental waves. Finally, Chapter 7 introduces a new method of inducing atmospheric refractivity fluctuations using coherent acoustic waves. The distribution of the artificial atmospheric refractive index is quantitatively calculated, and the feasibility of purposefully affecting radio wave propagation is demonstrated via experiments. The potential applications of synthetically controlling the radio wave propagation by the artificial refractivity fluctuation structure are explained, and future research directions are summarized.

Mahmut Reyhanoglu, PhD
University of North Carolina Asheville,
Mechatronics Engineering Laboratory,
Asheville, North Carolina,
USA

Section 1

Computational Studies of Acoustic Waves

Chapter 1

Optimized Finite Difference Methods for Seismic Acoustic Wave Modeling

Yanfei Wang and Wenquan Liang

Additional information is available at the end of the chapter

http://dx.doi.org/10.5772/intechopen.71647

Abstract

The finite difference (FD) methods are widely used for approximating the partial derivatives in the acoustic/elastic wave equation. Grid dispersion is one of the key numerical problems and will directly influence the accuracy of the result because of the discretization of the partial derivatives in the wave equation. Therefore, it is of great importance to suppress the grid dispersion by optimizing the FD coefficient. Various optimized methods are introduced in this chapter to determine the FD coefficient. Usually, the identical staggered grid finite difference operator is used for all of the first-order spatial derivatives in the first-order wave equation. In this chapter, we introduce a new staggered grid FD scheme which can improve the efficiency while still preserving high accuracy for the first-order acoustic/elastic wave equation modeling. It uses different staggered grid FD operators for different spatial derivatives in the first-order wave equation. The staggered grid FD coefficients of the new FD scheme can be obtained with a linear method. At last, numerical experiments were done to demonstrate the effectiveness of the introduced method.

Keywords: finite difference scheme, optimized finite difference coefficient, staggered grid, regularization, wave equation

1. Introduction

The propagation of seismic waves through the Earth's subsurface is described by the wave equation, one of the partial differential equations (PDEs), which describe many of the fundamental natural laws. When the subsurface earth structure is complex, it is difficult to obtain the analytic results. The finite difference (FD) method is one of most widely used numerical methods for wave equation modeling because of its high efficiency, smaller memory requirement, and easy implementation [1–7].

The first application of the FD method to wave equation modeling can be possibly traced back to Alterman and Karal [1]. Alford et al. took the grid dispersion analysis for the second-order and fourth-order FD operators and stated that it is necessary to use high-order FD operators [8]. Kelly et al. further illustrated the grid dispersion, absorbing boundary condition, and other implementation aspects of the FD method [9].

Madariaga developed a staggered grid FD scheme to solve a rupture propagation problem [10]. Virieux adapted this scheme to elastic SH waves and P-SV waves in a 2D Cartesian system [2, 3]. Levander introduced a fourth-order staggered grid FD operator in the space domain to improve accuracy [11].

Grid dispersion is one of the key numerical problems affecting practical usage when utilizing the FD method. Since the traditional FD coefficient obtained in the space domain with the Taylor expansion method is only accurate for a very limited wavenumber range [4], many efforts are paid to reducing the grid dispersion with optimized FD coefficient. Yang et al. proposed the nearly analytic discrete method for wave equation and later improved this method [12, 13]. Chen proposed high-order time discretization method to reduce the dispersion caused by the temporal discretization [14, 15]. The Fourier FD was introduced by Song and Fomel with the combination of fast Fourier transform and finite difference operators [16]. Chu and Stoffa improved the FD methods with a scaled binomial windowed FD scheme that leads to more precise discrete operators [17]. Fomel et al. introduced low-rank approximation of the wave propagator matrix to reduce the cost of wave extrapolation [18].

Generally, the FD coefficients of the spatial derivative are determined only in the spatial domain. However, wave equations are solved in the temporal and spatial domains simultaneously. Finkelstein and Kastner propose a systematic design methodology for obtaining FD coefficients to reduce dispersion, which allows the exact phase velocity or (and) group velocity dispersion relationship to be satisfied at some designated frequencies in the temporal-spatial domain [19, 20]. Etgen proposed minimizing the phase velocity error using the least squares (LS) method [21]. Liu and Sen propose a new time-space domain method to determine the higher order FD coefficients for 1D, 2D, and 3D wave equations [22], and then they use this method to get the staggered grid FD coefficients [23]. Zhang and Yao proposed the use of the simulated annealing algorithm and gave an error limitation for determining the FD coefficients in the space or the time-space domain [24]. Liang et al. proposed utilizing the linear method to determine the FD coefficient in the time-space domain [25]. Ren and Liu developed a novel optimal time-space domain staggered grid FD scheme and used least squares method to get the FD coefficients [26]. Wang et al. proposed the regularized optimization method to get the staggered grid FD coefficient in the time-space domain [27]. Chen et al. used K space operator-based high-order staggered grid FD method to improve accuracy [28]. Yong et al. proposed using the optimized equivalent staggered grid FD method with three sets of FD coefficients to improve the simulation accuracy [29]. Compared with the traditional high-order staggered grid FD coefficient obtained by the Taylor expansion method, these methods greatly improved the accuracy with the optimized FD coefficient.

Another way to improve the accuracy and efficiency of the FD methods is using new FD stencil. Liu and Sen studied the rhombus stencil and found that it can reach high-order accuracy along all directions [30]. Liu et al. formulated an explicit time evolution scheme with high temporal accuracy by using a new FD stencil for the second-order wave equation [31]. Tan and Huang propose a staggered grid FD stencil with added points in the diagonal direction for the first-order wave equation [32, 33]. Compared with the traditional staggered grid FD stencil, these methods improved the efficiency by using a larger time step while still preserving high accuracy.

2. Acoustic/elastic wave equations

The first-order velocity-stress acoustic wave equation can be described as

$$\frac{\partial P}{\partial t} = -v^2\left(\frac{\partial v_z}{\partial z} + \frac{\partial v_x}{\partial x}\right), \tag{1}$$

$$\frac{\partial v_x}{\partial t} = -\frac{\partial P}{\partial x}, \tag{2}$$

$$\frac{\partial v_z}{\partial t} = -\frac{\partial P}{\partial z}. \tag{3}$$

where P is the acoustic pressure fluctuation, v is the wave propagation speed, and v_x and v_z are the particle velocities.

Substituting Eqs. (2) and (3) into Eq. (1), the second-order acoustic wave equation can be written as

$$\frac{\partial^2 p}{\partial x^2} + \frac{\partial^2 p}{\partial z^2} = \frac{1}{v^2}\frac{\partial^2 p}{\partial t^2} \tag{4}$$

The first-order elastic wave equations in 2D heterogeneous media are [3]

$$\frac{\partial v_x}{\partial t} = \frac{\partial \tau_{xx}}{\partial x} + \frac{\partial \tau_{xz}}{\partial z}, \tag{5}$$

$$\frac{\partial v_z}{\partial t} = \frac{\partial \tau_{xz}}{\partial x} + \frac{\partial \tau_{zz}}{\partial z}, \tag{6}$$

$$\frac{\partial \tau_{xx}}{\partial t} = \alpha^2\frac{\partial v_x}{\partial x} + \left(\alpha^2 - 2\beta^2\right)\frac{\partial v_z}{\partial z}, \tag{7}$$

$$\frac{\partial \tau_{zz}}{\partial t} = \alpha^2\frac{\partial v_z}{\partial z} + \left(\alpha^2 - 2\beta^2\right)\frac{\partial v_x}{\partial x}, \tag{8}$$

$$\frac{\partial \tau_{xz}}{\partial t} = \beta^2\left(\frac{\partial v_x}{\partial z} + \frac{\partial v_z}{\partial x}\right). \tag{9}$$

where (v_x, v_z) is the velocity vector,$(\tau_{xx}, \tau_{zz}, \tau_{xz})$ is the stress vector, and α and β are the P- and S wave propagation speeds, respectively.

Substituting Eqs. (7)–(9) into Eqs. (5)–(6), the second-order elastic wave equation can be written as

$$\frac{\partial^2 v_x}{\partial t^2} = \alpha^2 \frac{\partial^2 v_x}{\partial x^2} + \left(\alpha^2 - \beta^2\right) \frac{\partial^2 v_z}{\partial x \partial z} + \beta^2 \frac{\partial^2 v_x}{\partial z^2}, \tag{10}$$

$$\frac{\partial^2 v_z}{\partial t^2} = \alpha^2 \frac{\partial^2 v_z}{\partial z^2} + \left(\alpha^2 - \beta^2\right) \frac{\partial^2 v_x}{\partial x \partial z} + \beta^2 \frac{\partial^2 v_z}{\partial x^2}. \tag{11}$$

3. Finite difference operators

The commonly used staggered grid FD scheme for the first-order acoustic wave equation is as follows:

$$\frac{\partial P}{\partial t} = \frac{-v^2}{h} \left(\sum_{m=1}^{M_1} c_m \left(v_{z_{0,m-1/2}} - v_{z_{0,-m+1/2}} \right) + \sum_{m=1}^{M_1} c_m \left(v_{x_{m-1/2,0}} - v_{x_{-m+1/2,0}} \right) \right), \tag{12}$$

$$\frac{\partial v_x}{\partial t} = \frac{-1}{h} \sum_{m=1}^{M_2} c_m \left(P^0_{m-1/2,0} - P^0_{-m+1/2,0} \right), \tag{13}$$

$$\frac{\partial v_z}{\partial t} = \frac{-1}{h} \sum_{m=1}^{M_2} c_m \left(P^0_{0,m-1/2} - P^0_{0,-m+1/2} \right), \tag{14}$$

$$Q^n_{m,j} = Q(x + mh, z + jh, t + n\tau); Q = v_x, v_z, P \tag{15}$$

where M_1 and M_2 are the length of the FD operators, c_m is the staggered grid FD coefficients, and h is the spatial grid interval.

The second-order FD operator is usually used for the first-order time derivative:

$$\frac{\partial P}{\partial t} = \frac{1}{\Delta t} \left(P^1_{0,0} - P^0_{0,0} \right). \tag{16}$$

where Δt is the time step.

The commonly used staggered grid FD scheme for the first-order elastic wave equation is as follows:

$$\frac{\partial v_x}{\partial t} = \frac{1}{h} \sum_{m=1}^{M_1} c_m \left[\tau_{xx}{}^0_{m-1/2,0} - \tau_{xx}{}^0_{-m+1/2,0} \right] + \frac{1}{h} \sum_{m=1}^{M_1} c_m \left[\tau_{xz}{}^0_{0,m-1/2} - \tau_{xz}{}^0_{0,-m+1/2} \right] \tag{17}$$

$$\frac{\partial v_z}{\partial t} = \frac{1}{h} \sum_{m=1}^{M_1} c_m \left[\tau_{xz}{}^0_{m-1/2,0} - \tau_{xz}{}^0_{-m+1/2,0} \right] + \frac{1}{h} \sum_{m=1}^{M_1} c_m \left[\tau_{zz}{}^0_{0,m-1/2} - \tau_{zz}{}^0_{0,-m+1/2} \right] \tag{18}$$

$$\frac{\partial \tau_{xx}}{\partial t} = \frac{\alpha^2}{h} \sum_{m=1}^{M_2} c_m \left[v_x{}^0_{m-1/2,0} - v_x{}^0_{-m+1/2,0} \right] + \frac{\alpha^2 - 2\beta^2}{h} \sum_{m=1}^{M_2} c_m \left[v_z{}^0_{0,m-1/2} - v_z{}^0_{0,-m+1/2} \right] \tag{19}$$

$$\frac{\partial \tau_{zz}}{\partial t} = \frac{\alpha^2 - 2\beta^2}{h}\sum_{m=1}^{M_2} c_m \left[{v_x}^0_{m-1/2,0} - {v_x}^0_{-m+1/2,0} \right] + \frac{\alpha^2}{h}\sum_{m=1}^{M_2} c_m \left[{v_z}^0_{0,m-1/2} - {v_z}^0_{0,-m+1/2} \right] \tag{20}$$

$$\frac{\partial \tau_{xz}}{\partial t} = \frac{\beta^2}{h}\sum_{m=1}^{M_2} c_m \left[{v_z}^0_{m-1/2,0} - {v_z}^0_{-m+1/2,0} \right] + \frac{\beta^2}{h}\sum_{m=1}^{M_2} c_m \left[{v_x}^0_{0,m-1/2} - {v_x}^0_{0,-m+1/2} \right] \tag{21}$$

$$Q^n_{m,j} = Q(x + mh, z + jh, t + n\tau); Q = v_x, v_z, \tau_{xx}, \tau_{zz}, \tau_{xz} \tag{22}$$

where M_1 and M_2 are the length of the FD operators, c_m is the staggered grid FD coefficient to be determined, and h is the spatial grid interval.

4. Optimizing finite difference operators

4.1. Optimizing finite difference operators for the acoustic wave equation

Using the plane wave theory, let

$$P^j_{m,n} = e^{i[k_x(x+mh)+k_z(z+jh)-\omega(t+n\tau)]}. \tag{23}$$

The following dispersion relation can be obtained by substituting Eqs. (13)–(14) into Eq. (12) [23, 26, 27]:

$$\left[\sum_{m=1}^{M} c_m \sin((m-0.5)k_x h)\right]^2 + \left[\sum_{m=1}^{M} c_m \sin((m-0.5)k_z h)\right]^2 \approx \frac{1}{2r^2}[1 - \cos(kv\tau)]. \tag{24}$$

where $r = v\Delta t/h$, $M_1 = M_2 = M$, and $(k_x, k_z) = k(\cos\theta, \sin\theta)$. It can be observed from Eq. (24) that the dispersion relation is complex and optimized methods are needed to address this problem.

Let c be the vector form of the FD coefficients, and denote the left side of Eq. 24 by [27]:

$$F(c) = \sum_{\theta=0}^{2\pi}\left\{\left[\sum_{m=1}^{M} c_m \sin((m-0.5)k_x h)\right]^2 + \left[\sum_{m=1}^{M} c_m \sin((m-0.5)k_z h)\right]^2\right\} \tag{25}$$

and the right side of Eq. 24 by

$$d = \sum_{\theta=0}^{2\pi}\frac{1}{2r^2}[1 - \cos(kv\tau)]. \tag{26}$$

The aim is to minimize the dispersion error for a fixed range of wavenumbers:

$$\Phi(c) = \sum_{k=0}^{\tilde{K}} [F(c) - d]^2 \to \min \tag{27}$$

The upper limit of the wavenumber range used for calculating the FD coefficients is based on the source frequency, the space grid interval, and the wave propagation speed [25]:

$$Ratio = \frac{\tilde{K}}{K_{total}} = \frac{2\pi f_{\max}/v}{\pi/h} = \frac{f_{\max}}{v/2h}. \tag{28}$$

The direct minimization of the objective function Φ for the FD coefficient may lead to unstable results. Therefore, regularizing technique was applied to restore stability. The regularization model is established as

$$J^{\alpha}(c) = \Phi(c) + \frac{1}{2}\alpha\|Dc\|^2, \tag{29}$$

where $\alpha > 0$ is a user-defined regularization parameter and D is a scale operator. The new task is the minimization of $J(c)$, and then the regularized optimization staggered grid FD coefficient can be obtained.

Another way to improve the efficiency and accuracy of the staggered grid FD methods is the utilization of the new staggered grid FD scheme. Different with the previous staggered grid FD scheme, the simplest centered second-order staggered grid FD operator can be used for the spatial derivatives in Eqs. (2) and (3), for example,

$$\frac{\partial P}{\partial t} = \frac{-v^2}{h}\left(\sum_{m=1}^{M} c_m\left(v_{z_{0,m-1/2}} - v_{z_{0,-m+1/2}}\right) + \sum_{m=1}^{M} c_m\left(v_{x_{m-1/2,0}} - v_{x_{-m+1/2,0}}\right)\right), \tag{30}$$

$$\frac{\partial v_x}{\partial t} = \frac{-\left(P^0_{1/2,0} - P^0_{-1/2,0}\right)}{h}, \tag{31}$$

$$\frac{\partial v_z}{\partial t} = \frac{-\left(P^0_{0,1/2} - P^0_{0,-1/2}\right)}{h}. \tag{32}$$

The staggered grid FD scheme in Eqs. (30)–(32) can be seen as a new staggered grid FD scheme for the first-order acoustic wave equation. The new staggered grid FD scheme is exactly the same as the traditional staggered grid FD scheme except if the staggered grid FD operator length is shorter for Eqs. (31) and (32). By carefully comparing Eqs. (12) and (14) with Eqs. (30) and (32), we find that the new staggered grid FD scheme is more efficient and can save about 45% of simulation time when M equals 7. It looks like the particle velocities v_x and v_z in Eqs. (31) and (32) are inaccurate since only the second-order staggered grid FD operators are used. However, this is not true since the staggered grid FD coefficient in Eq. (30) is optimized with Eqs. (31) and (32) in consideration. In the following, the huge advantage of the new staggered grid FD scheme will be demonstrated because it can reduce the simulation time while still preserving high accuracy compared with the traditional staggered grid FD scheme.

To get the staggered grid FD coefficient in Eq. (30), we substitute Eqs. (31) and (32) into Eq. (30), using the plane wave theory. Then we get

$$\begin{aligned}&\left(e^{ik_z\frac{h}{2}}-e^{-ik_z\frac{h}{2}}\right)\sum_{m=1}^{M}c_m\left(e^{ik_z\left(m-\frac{1}{2}\right)h}-e^{-ik_z\left(m-\frac{1}{2}\right)h}\right)\\&+\left(e^{ik_x\frac{h}{2}}-e^{-ik_x\frac{h}{2}}\right)\sum_{m=1}^{M}c_m\left(e^{ik_x\left(m-\frac{1}{2}\right)h}-e^{-ik_x\left(m-\frac{1}{2}\right)h}\right)=\frac{h^2}{v^2}\frac{e^{i\omega\tau}+e^{-i\omega\tau}-2}{\Delta t^2}\end{aligned}\tag{33}$$

From Eq. (33), the following dispersion relation can be obtained in the frequency-wavenumber domain (it is a special case of Eq. (24)):

$$-2\sin\frac{k_zh}{2}\sum_{m=1}^{M}c_m(\sin(m-0.5)k_zh)-2\sin\frac{k_xh}{2}\sum_{m=1}^{M}c_m(\sin(m-0.5)k_xh)=r^{-2}[\cos(\omega\tau)-1].\tag{34}$$

Using the basic trigonometric function

$$\sin\alpha\sin\beta=\frac{-[\cos(\alpha+\beta)-\cos(\alpha-\beta)]}{2},\tag{35}$$

we obtain Eq. (36) from Eq. (34):

$$\sum_{m=1}^{M}c_m[\cos(mk_xh)-\cos((m-1)k_xh)+\cos(mk_zh)-\cos((m-1)k_zh)]=r^{-2}[\cos(\omega\tau)-1].\tag{36}$$

Similarly, the new dispersion relation for the 3D first-order acoustic wave equation in the frequency-wavenumber domain is

$$\begin{aligned}&\sum_{m=1}^{M}c_m\left[\cos(mk_xh)-\cos((m-1)k_xh)+\cos\left(mk_yh\right)-\cos\left((m-1)k_yh\right)+\cos(mk_zh)-\cos((m-1)k_zh)\right]\\&=r^{-2}[\cos(\omega\tau)-1]\end{aligned}\tag{37}$$

where $\left(k_x,k_y,k_z\right)=k\left(\sin\theta\cos\phi,\sin\theta\sin\phi,\cos\theta\right)$.

Compared with the traditional dispersion relation in Eq. (24), the new dispersion relation in Eqs. (36) and (37) is linear and much simpler.

We assume that there are M equally distributed wavenumber points satisfying the dispersion relation within the wavenumber range specified by Eq. (28). Then, we establish the linear equation from Eq. (37) for the 3D case [25]:

$$\sum_{\phi=0}^{2\pi}\sum_{\theta=0}^{2\pi}\begin{pmatrix} a^h_{k_{1,x},1}+a^h_{k_{1,y},1}+a^h_{k_{1,z},1} & \cdots & a^h_{k_{1,x},M}+a^h_{k_{1,y},M}+a^h_{k_{1,x},M} \\ \vdots & & \vdots \\ a^h_{k_{M,x},1}+a^h_{k_{M,y},1}+a^h_{k_{M,z},1} & \cdots & a^h_{k_{M,x},M}+a^h_{k_{M,y},M}+a^h_{k_{M,x},M} \end{pmatrix}\begin{pmatrix} c_1 \\ \vdots \\ c_M \end{pmatrix}$$
$$=\frac{1}{r^2}\sum_{\phi=0}^{2\pi}\sum_{\theta=0}^{2\pi}\begin{pmatrix} \cos(k_1 v\tau)-1 \\ \vdots \\ \cos(k_M v\tau)-1 \end{pmatrix} \tag{38}$$

where $a^h_{k_l,m} = \cos(mk_l h) - \cos((m-1)k_l h)$, the ith component of $k_l(l=x,y,z)$ is represented as $k_{i,l}$, $k_x = k\cos\theta\cos\phi, k_y = k\cos\theta\sin\phi, k_z = k\sin\theta$, and $k(i)$ for each $i = 1,2,\ldots,M+1$ is equally distributed between 0 and $Ratio \times \pi / h$, where $Ratio$ is determined by Eq. (28). In the following, we will demonstrate that the new staggered grid FD scheme in Eqs. (30)–(32) has similar accuracy compared with the computational intensive traditional staggered grid FD scheme in Eqs. (12)–(14).

The 2D dispersion error δ of the new staggered grid FD scheme is defined as

$$\delta = \frac{v_{FD}}{v} = \frac{1}{rkh}\arccos\left(1 + r^2 q_2\right). \tag{39}$$

where

$$q_2 = \sum_{m=1}^{M} c_m[\cos(mk_x h) - \cos((m-1)k_x h) + \cos(mk_z h) - \cos((m-1)k_z h)] \tag{40}$$

The difference between the FD propagation time and the exact propagation time through one grid is defined as [23]

$$\varepsilon = \frac{h}{v_{FD}} - \frac{h}{v} = \frac{h}{v}\left(\frac{v}{v_{FD}} - 1\right) = \frac{h}{v}\left(\frac{1}{\delta} - 1\right) \tag{41}$$

Figures 1 and **2** show the dispersion error curves of the traditional and the new staggered grid FD schemes for the homogeneous acoustic model in 2D. All the FD coefficients are determined in the time-space domain with M = 7. From **Figures 1** and **2**, we get the conclusion that the new staggered grid FD scheme can also preserve the dispersion relation in a pretty wider range compared with the traditional staggered grid FD methods. For example, with $r = 0.0075$ in the 2D case, both of them can preserve the dispersion error under 10^{-5} within 80% of kh range. However, the new staggered grid FD scheme saves wave equation simulation time because Eqs. (31) and (32) are much simpler than Eqs. (13) and (14).

Let the left part of Eq. (36) as.

$$g = \sum_{m=1}^{M} c_m[\cos(mk_x h) - \cos((m-1)k_x h) + \cos(mk_z h) - \cos((m-1)k_z h)]. \tag{42}$$

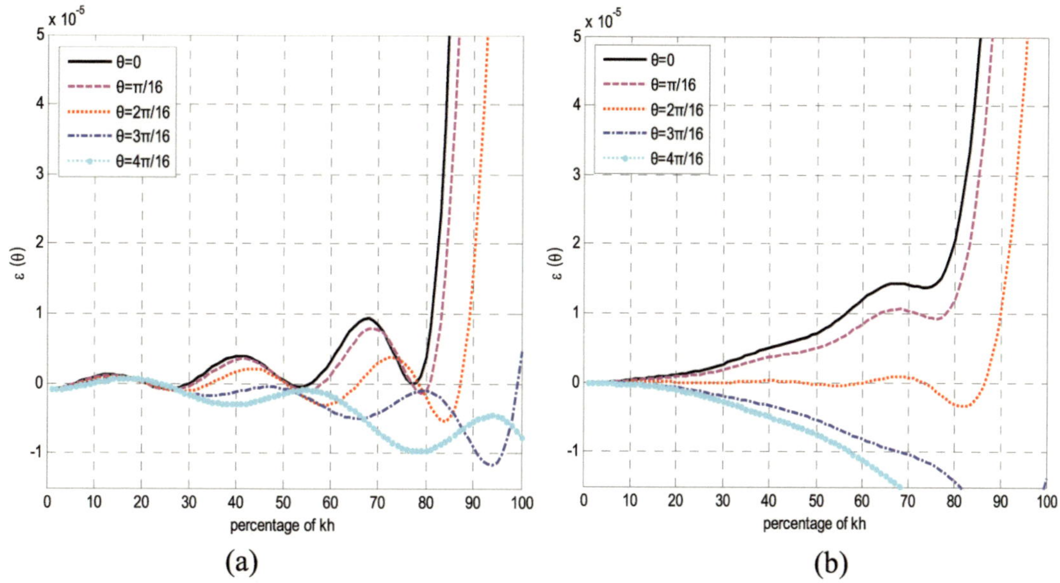

Figure 1. Dispersion error curves of the traditional staggered grid FD schemes. (a) $r = 0.075$ and (b) $r = 0.225$.

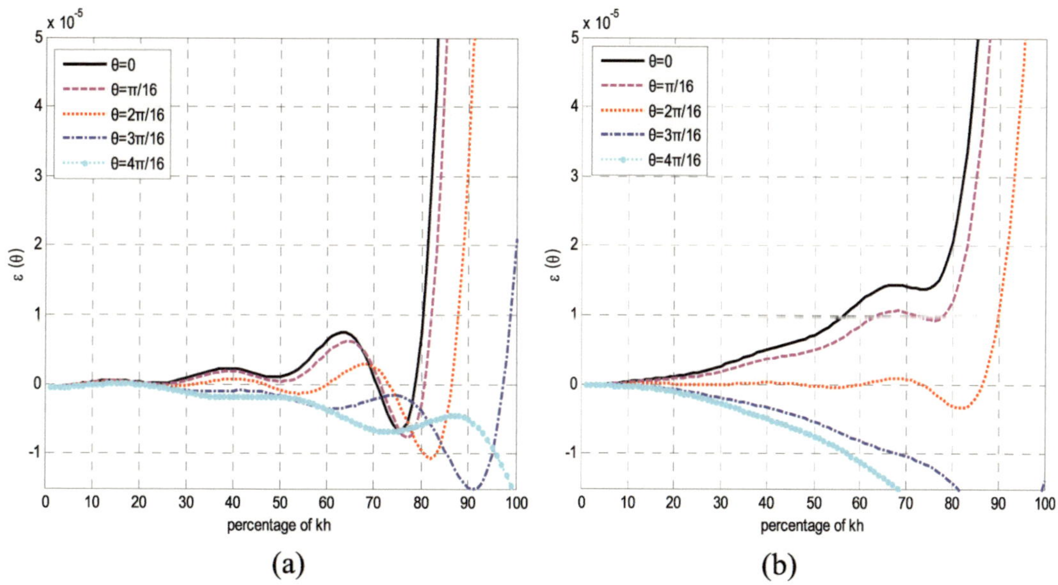

Figure 2. Dispersion error curves of the new staggered grid FD schemes. (a) $r = 0.075$ and (b) $r = 0.225$.

From dispersion relation Eq. (36), it is obvious that.

$$g < 0;\ \ r^2 g \geq -2. \tag{43}$$

Then, the stability condition of the new staggered grid FD scheme is (from Eq. (36) with $kh = \pi$).

$$r \le \sqrt{-2/g} \le \sqrt{\frac{-2}{\sum_{m=1}^{M} 4c_m(-1)^m}} = \frac{1}{\sqrt{2}}\sqrt{\sum_{m=1}^{M} c_m}^{\,-1}. \tag{44}$$

Figure 3 shows the stability condition of the traditional and the new staggered grid FD scheme in 2D. We can see that the stability condition becomes stricter with the increase of the FD operator length. It also shows that the new staggered grid FD scheme's stability condition is a little bit better than the previous staggered grid FD scheme. For example, the stability conditions are $r < 0.54$ and $r < 0.57$, respectively, for the traditional and the new staggered grid FD scheme with M = 7.

4.2. Optimizing finite difference operators for the elastic wave equation

Eqs. (10) and (11) can be written as [11].

$$\begin{pmatrix} \alpha^2 D_{xx} + \beta^2 D_{zz} - D_{tt} & (\alpha^2 - \beta^2) D_{xz} \\ (\alpha^2 - \beta^2) D_{xz} & \alpha^2 D_{zz} + \beta^2 D_{xx} - D_{tt} \end{pmatrix} \begin{pmatrix} v_x \\ v_z \end{pmatrix} = 0. \tag{45}$$

The two roots give the following dispersion relation [11]:

$$D_{tt} = \frac{1}{2}(\alpha^2 + \beta^2)(D_{xx} + D_{zz}) \pm \frac{1}{2}(\alpha^2 - \beta^2)\sqrt{(D_{xx} + D_{zz})^2 - 4(D_{xx}D_{zz} - D_{xz}D_{xz})}. \tag{46}$$

Suppose $D_{xx}D_{zz} = D_{xz}D_{xz}$, then two equations can be obtained from Eq. (46):

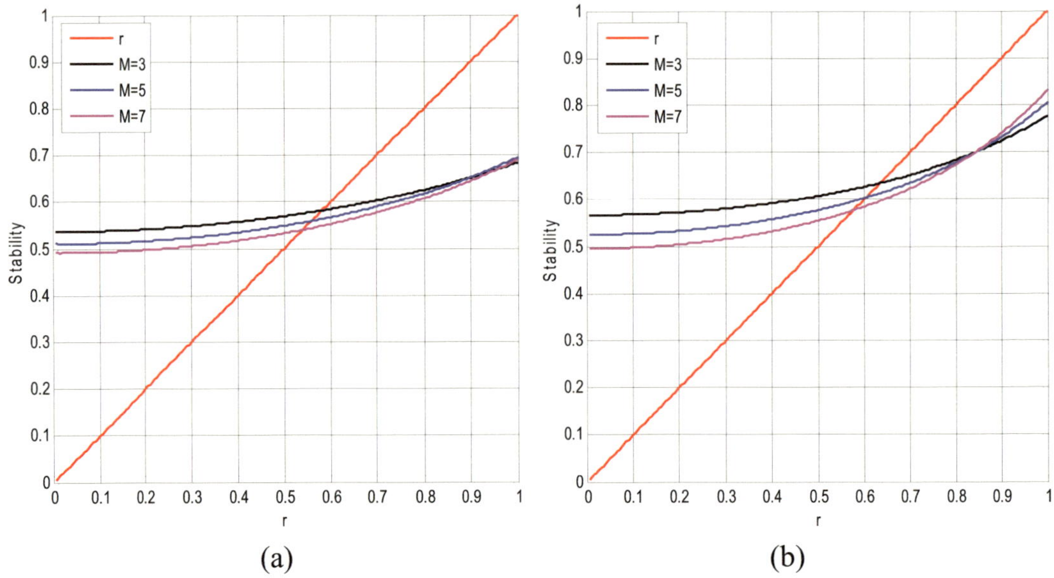

Figure 3. Stability condition in 2D. (a) The traditional staggered grid FD scheme and (b) the new staggered grid FD scheme.

$$\alpha^2 D_{xx} + \alpha^2 D_{zz} - D_{tt} = 0. \tag{47}$$

$$\beta^2 D_{xx} + \beta^2 D_{zz} - D_{tt} = 0. \tag{48}$$

Usually, Eq. (48) is used to determine the FD coefficient. For the first-order staggered grid FD scheme, the following dispersion relation can be obtained from Eq. (48):

$$\left[\sum_{m=1}^{M} c_m \sin\left((m-0.5)k_x h\right)\right]^2 + \left[\sum_{m=1}^{M} c_m \sin\left((m-0.5)k_z h\right)\right]^2 \approx \frac{1}{2r^2}\left[1 - \cos\left(kv\tau\right)\right]. \tag{49}$$

where $r = \beta\Delta t/h$, $M_1 = M_2 = M$, and $(k_x, k_z) = k(\cos\theta, \sin\theta)$. It can be observed from Eq. (49) that the dispersion relation is nonlinear and regularized optimized methods can address this problem similarly.

Different with previous staggered grid FD scheme for the first-order elastic wave equation, the simplest centered second-order staggered grid FD operator can be used for the spatial derivatives in Eqs. (7)–(9):

$$\frac{\partial v_x}{\partial t} = \frac{1}{h}\sum_{m=1}^{M_1} c_m \left[{\tau_{xx}}^0_{m-1/2,0} - {\tau_{xx}}^0_{-m+1/2,0}\right] + \frac{1}{h}\sum_{m=1}^{M_1} c_m \left[{\tau_{xz}}^0_{0,m-1/2} - {\tau_{xz}}^0_{0,-m+1/2}\right] \tag{50}$$

$$\frac{\partial v_z}{\partial t} = \frac{1}{h}\sum_{m=1}^{M_1} c_m \left[{\tau_{xz}}^0_{m-1/2,0} - {\tau_{xz}}^0_{-m+1/2,0}\right] + \frac{1}{h}\sum_{m=1}^{M_1} c_m \left[{\tau_{zz}}^0_{0,m-1/2} - {\tau_{zz}}^0_{0,-m+1/2}\right] \tag{51}$$

$$\frac{\partial \tau_{xx}}{\partial t} = \frac{\alpha^2}{h}\left[{v_x}^0_{1/2,0} - {v_x}^0_{-1/2,0}\right] + \frac{\alpha^2 - 2\beta^2}{h}\left[{v_z}^0_{0,1/2} - {v_z}^0_{0,-1/2}\right] \tag{52}$$

$$\frac{\partial \tau_{xx}}{\partial t} = \frac{\alpha^2 - 2\beta^2}{h}\left[{v_x}^0_{1/2,0} - {v_x}^0_{-1/2,0}\right] + \frac{\alpha^2}{h}\left[{v_z}^0_{0,1/2} - {v_z}^0_{0,-1/2}\right] \tag{53}$$

$$\frac{\partial \tau_{xz}}{\partial t} = \frac{\beta^2}{h}\left[{v_z}^0_{1/2,0} - {v_z}^0_{-1/2,0}\right] + \frac{\beta^2}{h}\left[{v_x}^0_{0,1/2} - {v_x}^0_{0,-1/2}\right] \tag{54}$$

The staggered grid FD scheme in Eqs. (52)–(54) is more efficient than the staggered grid FD scheme in Eqs. (19)–(21). It will be demonstrated later that the staggered grid FD scheme in Eqs. (52)–(54) is accurate for the stress vector $(\tau_{xx}, \tau_{zz}, \tau_{xz})$ even when only second-order staggered grid FD operator is used.

Then, the new dispersion relation can be obtained from Eq. (49) in the frequency-wavenumber domain:

$$\sum_{m=1}^{M_1} c_m\left[\cos\left(mk_x h\right) - \cos\left((m-1)k_x h\right) + \cos\left(mk_z h\right) - \cos\left((m-1)k_z h\right)\right] = r^{-2}\left[\cos\left(\omega\tau\right) - 1\right]. \tag{55}$$

The staggered grid FD coefficient can be obtained similarly using the linear method.

5. Experiments

5.1. Acoustic wave equation

5.1.1. Numerical modeling in the layered velocity model

We first consider a layered velocity model. The velocity is 1500 m/s for the first layer and 2500 m/s for the second layer as shown in **Figure 4**. The sponge boundary code in CREWES Toolbox is used to reduce artificial reflection waves [34]. A Ricker wavelet with the main frequency as 14.3 Hz was used as the seismic source. The seismic source position is denoted as a asterisk, and the receivers A and B are denoted as a circle and a diamond from top to bottom, respectively. The space grid interval is 20m, the FD operator length M is 7 and the time step is 1.5 ms. The staggered grid FD coefficients used in **Figure 4** are shown in **Table 1**.

The seismograms recorded at positions A and B by different methods are presented in **Figure 5**. **Figure 5(a)** is obtained with the traditional staggered grid FD scheme with the FD coefficient obtained in the space domain by Taylor expansion method [17]. The grid dispersion is obvious. **Figure 5(b)** is obtained with the traditional staggered grid FD scheme with the coefficient obtained in the space domain by the least squares method [35]. The staggered grid FD

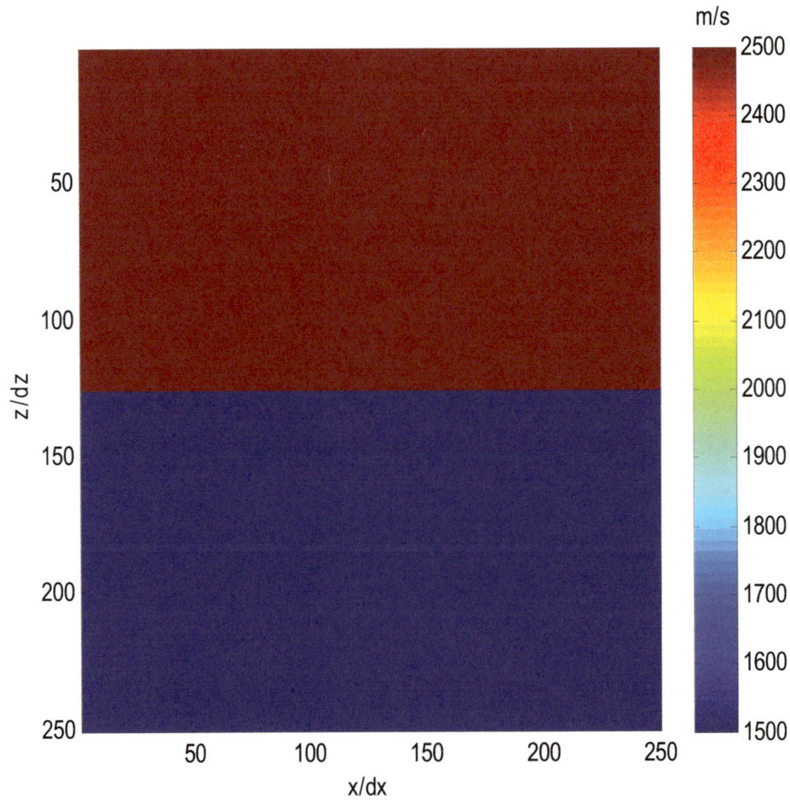

Figure 4. Velocity model.

	c_1	c_2	c_3	c_4	c_5	c_6	c_7
	1.22861	−0.102384	0.0204768	−0.00417893	0.000689454	−0.0000769225	0.00000423651
	1.25438	−0.1235307	0.03467231	−0.01192915	0.00405709	−0.001191005	0.0002263204
v = 1500 m/s	1.57866	−0.296598	0.0949307	−0.0344762	0.0120067	−0.00344529	0.000605554

Table 1. Staggered grid FD coefficient used to obtain the seismograms in **Figure 4** with the space grid interval equals 20 m, and the time step equals 1.5 ms. In the first row is the traditional staggered grid FD coefficient obtained from Table 3 of Chu and Stoffa [17]; in the second row is the least squares staggered grid FD coefficient obtained from Table 3 of Liu [35]; and in the last rows are the staggered grid FD coefficients used for Eq. (30). Eqs. (31) and (32) use the simplest second-order staggered grid FD operator.

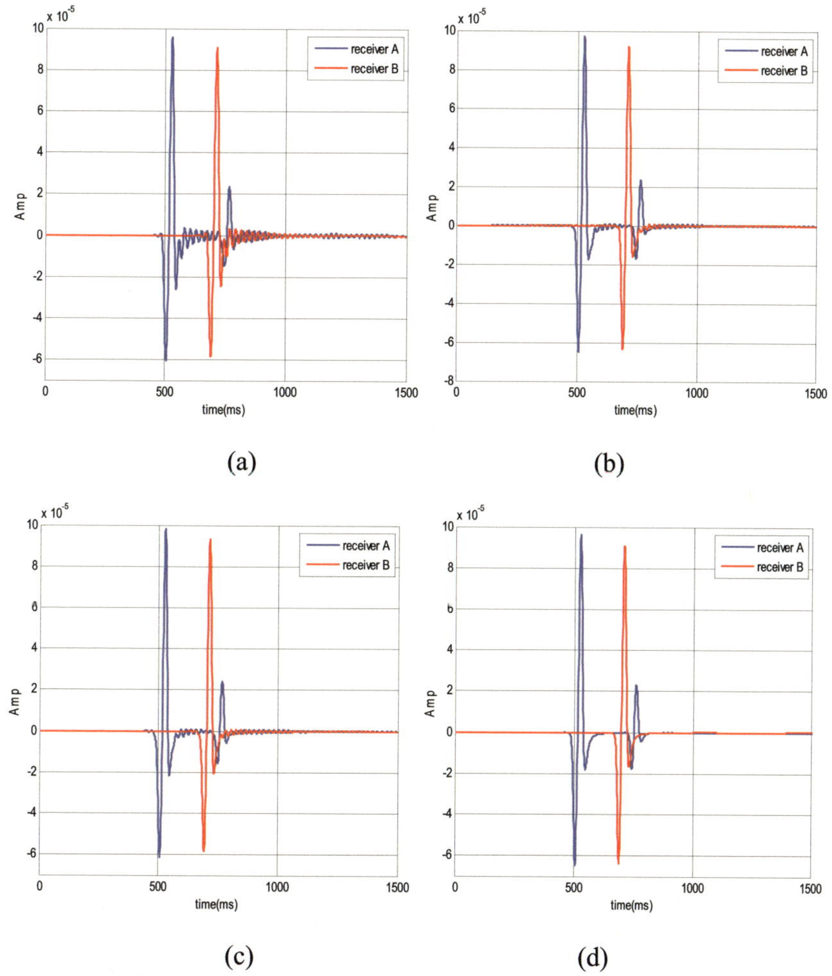

Figure 5. Seismograms recorded by different simulation methods. (a) The traditional staggered grid FD scheme with FD coefficients determined in the space domain by Taylor expansion method, (b) the traditional staggered grid FD scheme with FD coefficients determined in the space domain by least squares method, (c) the new staggered grid FD scheme with FD coefficients determined in the time-space domain by the linear method, and (d) the pseudo-spectrum method.

coefficient provided by Liu is one of the best staggered grid FD coefficient provided in recent years [35]. **Figure 5(c)** is obtained with the new staggered grid FD scheme with the coefficient determined in the time-space domain by the linear method. **Figure 5(d)** is obtained with the pseudo-spectrum method with the second-order acoustic wave equation. We observe that the grid dispersion in **Figure 5(b)** and **(c)** is similar to each other and is close to the nearly analytic results obtained with the pseudo-spectrum method in **Figure 5(d)**. However, the required simulation time is reduced by using the new staggered grid FD scheme because Eqs. (31) and (32) are much simpler than Eqs. (13) and (14).

5.1.2. Numerical modeling in the salt model

Figure 6 shows the salt model from Society of Exploration of geophysicists with variations of velocities from 1486 to 4790 m/s. The seismic source function is the same as the previous example. The spatial sampling interval is 20 m, temporal step is 1 ms, and $M = 7$ for the staggered grid FD operators in **Figure 7(a)** and **(b)**. In **Figure 7(c)**, the parameters are $M = 7$ for the spatial derivatives in Eq. (1), and $M = 1$ for the spatial derivatives in Eqs. (2) and (3). The pseudo-spectrum method is used for the second-order acoustic wave equation as shown **Figure 7(d)**.

Figure 7(a) is obtained the with the traditional staggered grid FD scheme with the coefficient obtained in the space domain by Taylor expansion method. The grid dispersion is obvious. **Figure 7(b)** is obtained with the traditional staggered grid FD scheme with the coefficient obtained in the time-space domain by the least squares method [27]. Most of the grid

Figure 6. SEG BP salt model.

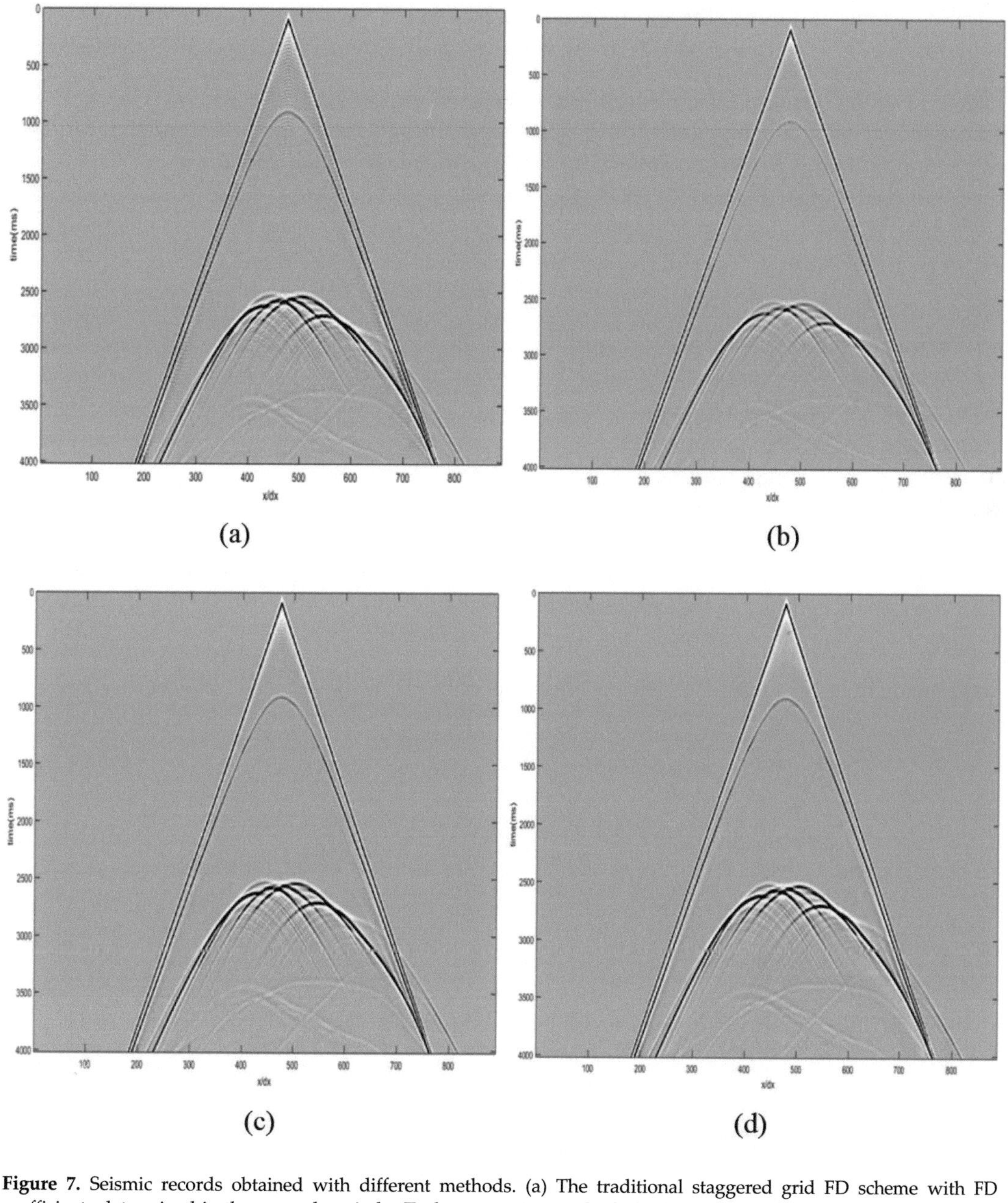

Figure 7. Seismic records obtained with different methods. (a) The traditional staggered grid FD scheme with FD coefficients determined in the space domain by Taylor expansion method, (b) the traditional staggered grid FD scheme with FD coefficients determined in the time-space domain by least squares method, (c) the new staggered grid FD scheme with FD coefficients determined in the time-space domain by the linear method, and (d) the pseudo-spectrum method.

dispersion is suppressed. **Figure 7(c)** is obtained the with the new staggered grid FD scheme with the coefficient obtained in the time-space domain by the linear method. The grid dispersion in **Figure 7(c)** is very similar to the grid dispersion in **Figure 7(b)**. However, the simulation time to get **Figure 7(c)** is reduced compared with the simulation time to get **Figure 7(b)**. Both seismic records in **Figure 7(b)** and **(c)** are close to seismic record in **Figure 7(d)**. We want to mention that the linear method is faster than the LS method to determine the FD coefficients.

Figure 8 Further compares the seismograms in **Figure 7** at position x/dx = 400. It is also observed that with the coefficient obtained in the space domain by Taylor expansion method, the grid dispersion is serious in the simulation result. The simulation results are almost overlapped for the traditional staggered grid FD scheme and new staggered grid FD scheme with optimized FD coefficient. However, the required simulation time is reduced by using the new staggered grid FD scheme because Eqs. (31) and (32) are much simpler than Eqs. (13) and (14).

Figure 9 compares snapshots of particle velocity v_x with the different staggered grid FD schemes at 2500 ms. it is also observed that with the coefficient obtained in the space domain by Taylor expansion method, the grid dispersion is most serious. The grid dispersion in **Figure 9(c)** is very similar to the grid dispersion in **Figure 9(b)**. It demonstrated that the new staggered grid FD scheme is accurate for the particle velocities in Eqs. (32) and (33) even when only second-order staggered grid FD operator is used.

5.2. Elastic wave equation

5.2.1. Numerical modeling in the homogeneous media

We first consider a homogeneous model. The P wave propagation speed is 2598 m/s, and the S wave velocity is 1500 m/s. The seismic source position is at the center of the model. The grid

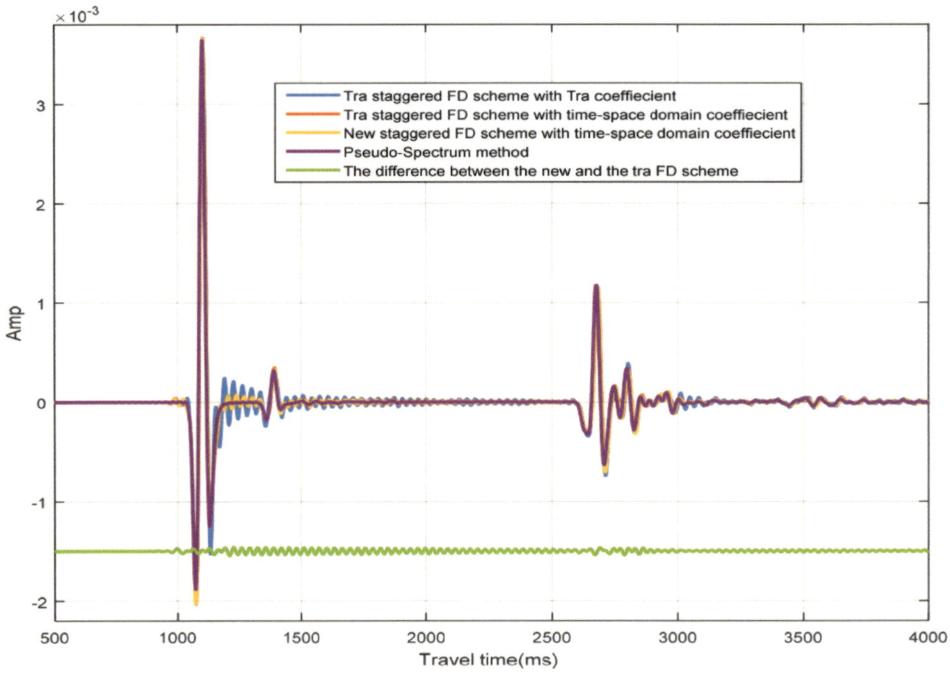

Figure 8. Seismograms at x/dx = 400 from **Figure 4(a)–(d)**.

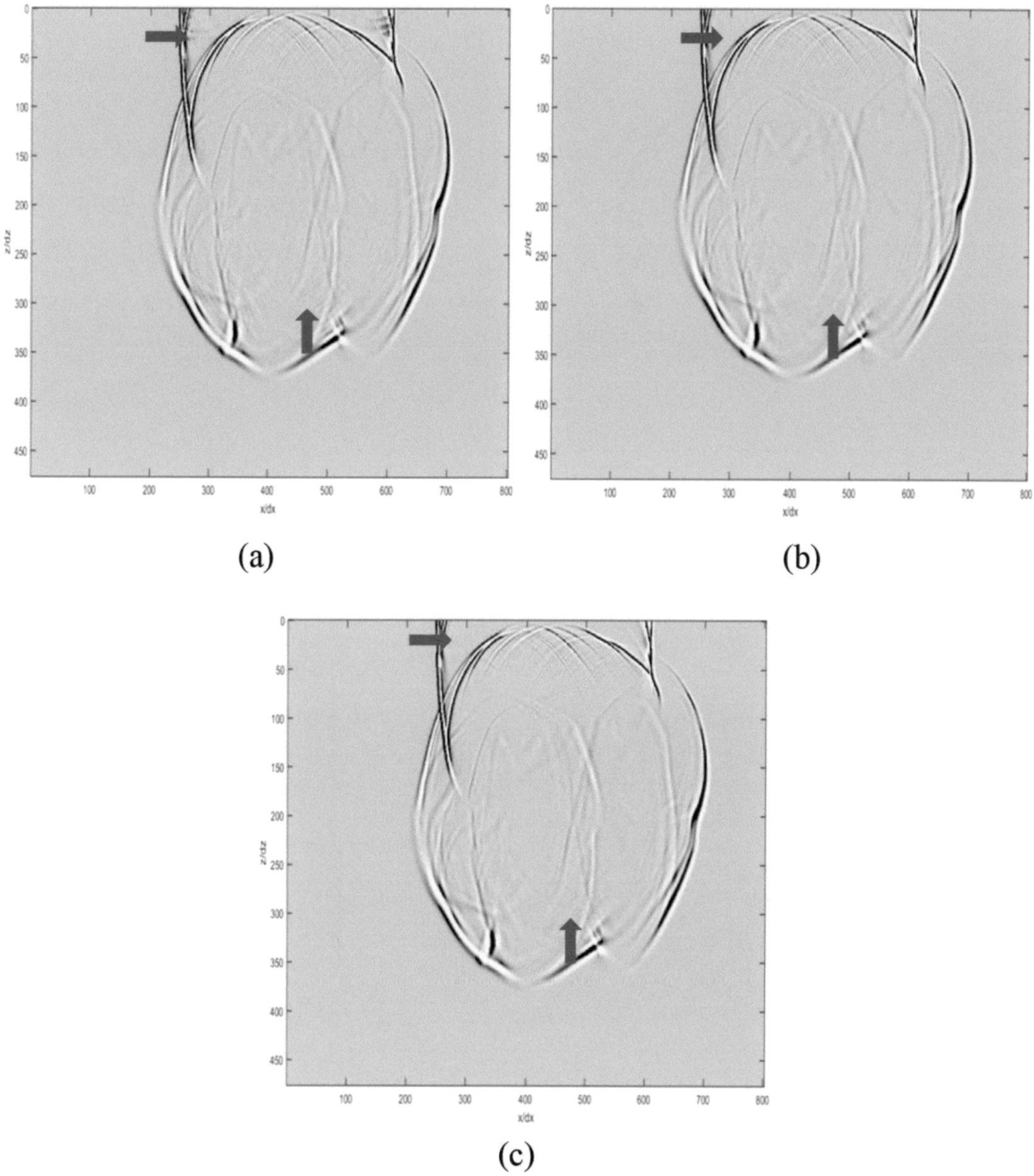

Figure 9. Particle velocity snapshots v_x obtained with different methods. (a) The traditional staggered grid FD scheme with the traditional FD coefficients, (b) the traditional staggered grid FD scheme with FD coefficients determined in the time-space domain by the least squares method, and (c) the new staggered grid FD scheme with FD coefficients determined in the time-space domain by the linear method.

space interval is 20 m, the time step is 1 ms, and the operator length M is 7. A Ricker wavelet with the main frequency as 14.3 Hz was used as the seismic source.

The snapshots of the horizontal component obtained by different staggered grid FD methods are presented in **Figure 10(a)–(c)**. **Figure 10(a)** is obtained with the traditional staggered grid

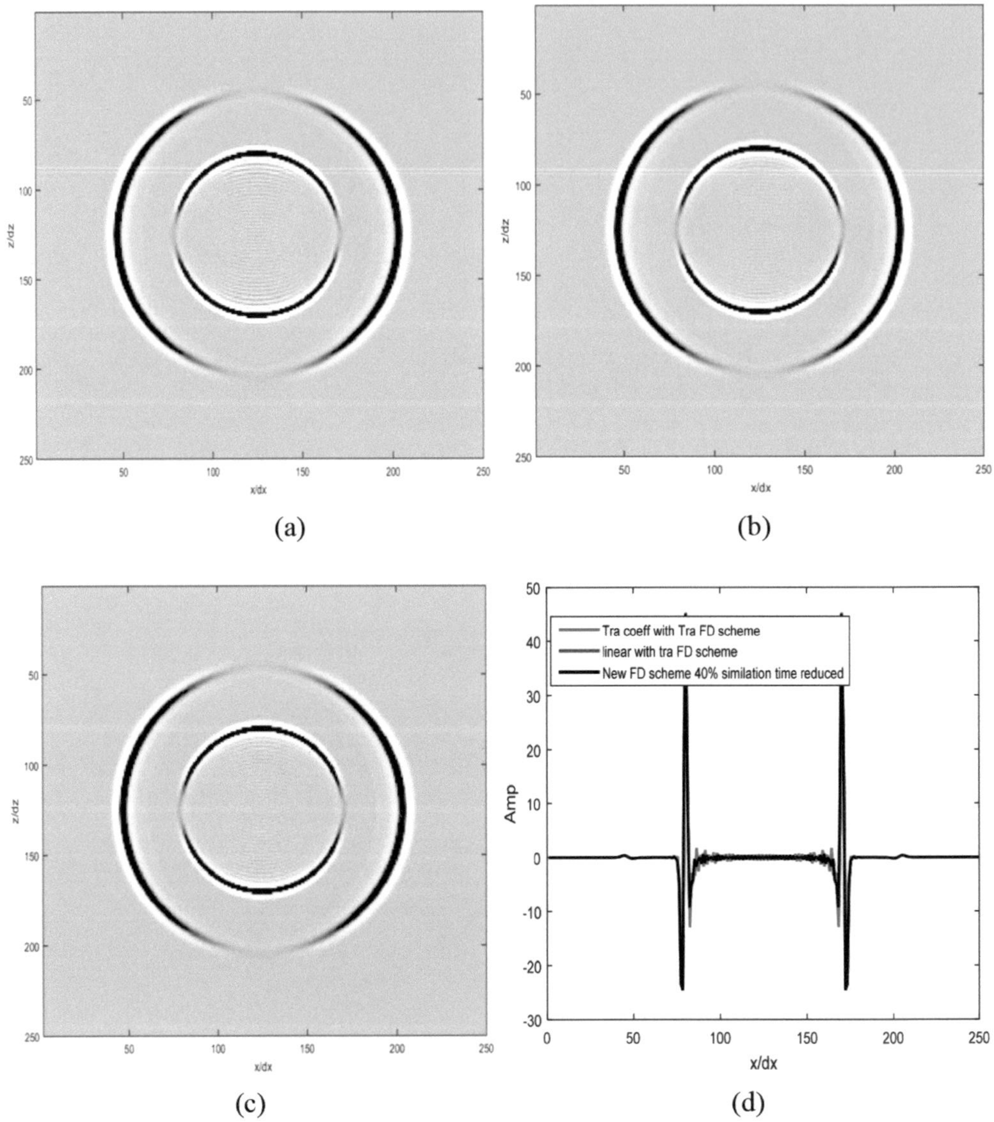

Figure 10. Snapshots and slices of snapshots of the horizontal component at 698 ms obtained by different simulation methods. (a) The traditional staggered gird FD scheme with traditional FD coefficient, (b) the traditional staggered gird FD scheme with new FD coefficient, (c) the new staggered grid FD scheme with new FD coefficient, and (d) slices of snapshots at x/dx = 125.

FD scheme with the traditional FD coefficient. The grid dispersion is obvious. **Figure 10(b)** is obtained with the traditional staggered grid FD scheme with the new FD coefficient. Compared with **Figure 10(a)**, the grid dispersion is suppressed. **Figure 10(c)** is obtained with the new staggered grid FD scheme. The grid dispersion curves in **Figure 10(b)** and **(c)** are very similar, which is further demonstrated in **Figure 10(d)**. However, with the new staggered grid FD scheme, we can save about 45% of the modeling time.

5.2.2. *Numerical modeling in the homogeneous media*

Figure 11 shows the salt model from Society of Exploration of geophysicists. The S wave velocity is obtained from the P wave velocity. The seismic source function is plotted as a red asterisk. The spatial sampling interval is 12.5 m, the temporal step is 1 ms, and $M = 7$ for staggered grid FD operators.

Figure 12 displays the seismic records of the horizontal component obtained by different staggered grid FD methods. **Figure 12(a)** is obtained with the traditional FD scheme with the traditional staggered grid FD coefficient. The grid dispersion is severe. **Figure 12(b)** is obtained with the traditional FD scheme with the staggered grid FD coefficient obtained by the least squares method. **Figure 12(c)** is obtained the with the new FD scheme with the staggered grid FD coefficient obtained by the linear method. It is observed that the grid dispersion in **Figure 12(b)** and **12(c)** is smaller than the grid dispersion in **Figure 12(a)**. **Figure 12(d)** is seismograms obtained from **Figure 12(a)–(c)**. It further demonstrated that the grid dispersion in **Figure 12(b)** and **(c)** is similar to each other and smaller than the grid dispersion in **Figure 12(a)**. However, with the new FD scheme, the simulation time is reduced about 45%. In our simulation, there are 525 grids in the z direction and 850 grids in the x direction. With the traditional

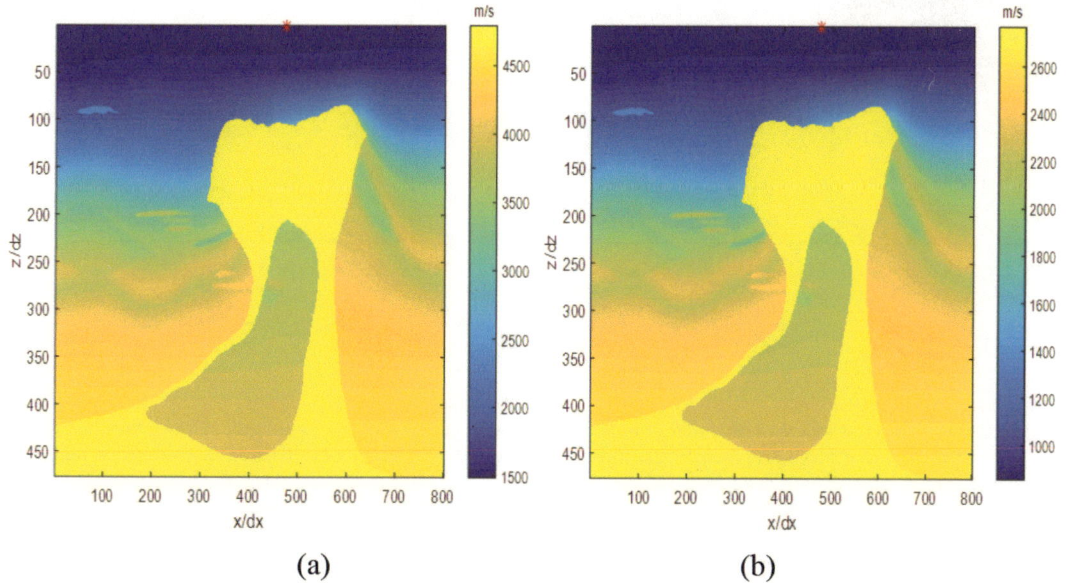

Figure 11. SEG BP salt model. (a) P wave velocity and (b) S wave velocity.

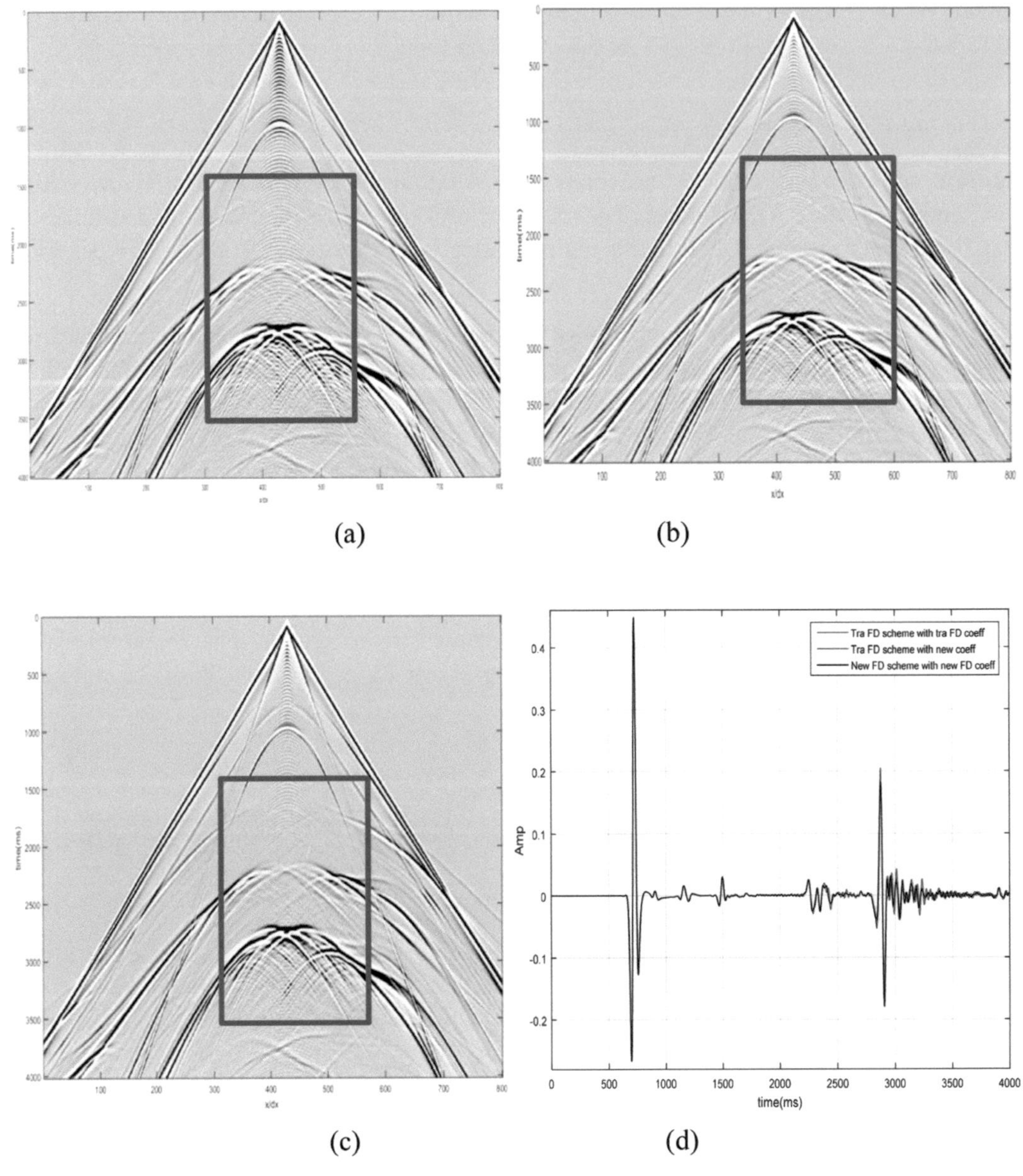

Figure 12. Seismic records of the horizontal component obtained with different staggered grid FD methods. (a) Seismic records obtained by traditional FD scheme with traditional staggered grid FD coefficient, (b) seismic records obtained by the traditional FD scheme with new staggered grid FD coefficient, (c) seismic records obtained by the new FD scheme with new staggered grid FD coefficient, and (d) seismograms obtained from (a) to (c) at position x/dx = 355.

FD scheme, the simulation time is 920 seconds. With the new staggered FD grid scheme, the simulation time is 530 seconds. The huge reduction in simulation time is due to using the shorter staggered FD operator for the spatial derivatives in Eqs. (7)–(9). **Figure 13** is the seismic records of the vertical component obtained by different FD methods. The same pattern can be observed from **Figure 13(a)–(d)**.

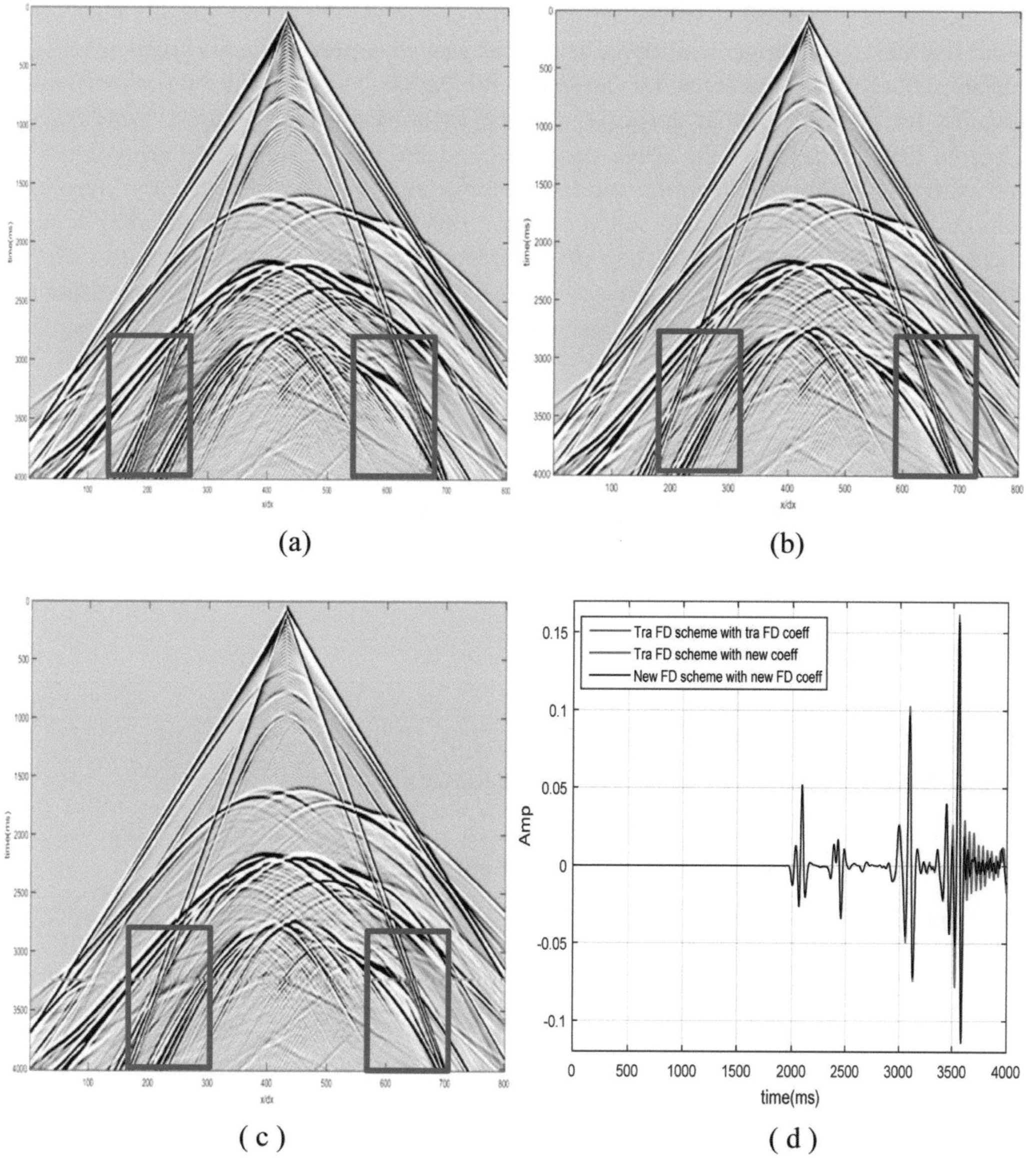

Figure 13. Seismic records of the vertical component obtained with different staggered grid FD methods. (a) The traditional FD scheme with traditional staggered grid FD coefficient, (b) the traditional FD scheme with new staggered grid FD coefficient, (c) the new FD scheme with new staggered grid FD coefficient, and (d) seismograms obtained from (a) to (c) at position x/dx = 200.

6. Discussion and conclusion

The FD method is the most commonly used numerical method for wave equation modeling. Suppressing the grid dispersion is an important research area. Optimization method is usually used to determine the FD coefficients which could preserve the dispersion relation in a wider range of wavenumber (Zhang and Yao [24]; Ren and Liu [26]; Tan and Huang [32, 33]). We introduced the regularized optimization method to determine the FD coefficient which would be more robust for extreme conditions. The other way to suppress the grid dispersion is the utilization of the new FD scheme for the spatial derivatives. We introduce to use different FD operators for different spatial derivatives in the first-order wave equation. With the new staggered grid FD scheme, the wave equation modeling speed was accelerated while still preserving high accuracy. Through numerical modeling, we conclude that the introduced methods are more efficient while still preserving high accuracy for the first-order acoustic/elastic wave equation modeling. As a result, the introduced methods can be a substitute for the traditional FD methods used in acoustic/elastic wave equation modeling, which are essential in forward seismic wave modeling and reverse-time migration.

Acknowledgements

This work is supported by the National Natural Science Foundation of China under grant numbers 41325016, 41704120, and 91630202.

Author details

Yanfei Wang[1,2,3]* and Wenquan Liang[4]

*Address all correspondence to: yfwang@mail.iggcas.ac.cn

1 Key Laboratory of Petroleum Resources Research, Institute of Geology and Geophysics, Chinese Academy of Sciences, Beijing, People's Republic of China

2 Institutions of Earth Science, Chinese Academy of Sciences, Beijing, People's Republic of China

3 University of Chinese Academy of Sciences, Beijing, People's Republic of China

4 College of Resource Engineering, Longyan University, Longyan, People's Republic of China

References

[1] Alterman Z, Karal FC. Propagation of elastic waves in layered media by finite difference methods. Bulletin of the Seismological Society of America. 1968;**58**(1):367-398

[2] Virieux J. SH-wave propagation in heterogeneous media: Velocity-stress finite- difference method. Geophysics. 1984;**49**(11):1933-1942

[3] Virieux J. P-SV wave propagation in heterogeneous media: Velocity-stress finite-difference method. Geophysics. 1986;**51**(4):889-901

[4] Dablain MA. The application of high-order differencing to the scalar wave equation. Geophysics. 1986;**51**(1):54-66

[5] Robertsson JO, Blanch JO, Symes WW. Viscoelastic finite-difference modeling. Geophysics. 1994;**59**(9):1444-1456

[6] Bohlen T, Wittkamp F. Three-dimensional viscoelastic time-domain finite-difference seismic modeling using the staggered Adams-Bashforth time integrator. Geophysical Journal International. 2016;**204**(3):1781-1788

[7] Etemadsaeed L, Moczo P, Kristek J, Ansari A, Kristekova M. A no-cost improved velocity-stress staggered-grid finite-difference scheme for modelling seismic wave propagation. Geophysical Journal International. 2016;**207**(1):481-511

[8] Alford RM, Kelly KR, Boore DM. Accuracy of finite-difference modeling of the acoustic wave equation. Geophysics. 1974;**39**(6):834-842

[9] Kelly KR, Ward RW, Treitel S, Alford RM. Synthetic seismograms: A finite-difference approach. Geophysics. 2012;**41**:2-27

[10] Madariaga R. Dynamics of an expanding circular fault. Bulletin of the Seismological Society of America. 1976;**66**(3):639-666

[11] Levander AR. Fourth-order finite-difference p-sv seismograms. Geophysics. 1988;**53**(11):1425-1436

[12] Yang D, Teng J, Zhang Z, Liu E. A nearly analytic discrete method for acoustic and elastic wave equations in anisotropic media. Bulletin of the Seismological Society of America. 2003;**93**(2):882-890

[13] Yang D, Lu M, Wu R, Peng J. An optimal nearly analytic discrete method for 2D acoustic and elastic wave equations. Bulletin of the Seismological Society of America. 2004;**94**(5):1982-1992

[14] Chen JB. High-order time discretizations in seismic modeling. Geophysics. 2007;**72**(5):SM115-SM122

[15] Chen JB. A stability formula for lax-Wendroff methods with fourth-order in time and general-order in space for the scalar wave equation. Geophysics. 2011;**76**(2):T37-T42

[16] Song X, Fomel S. Fourier finite-difference wave propagation. Geophysics. 2011;**76**(5):T123-T129

[17] Chu C, Stoffa PL. Determination of finite-difference weights using scaled binomial windows. Geophysics. 2012;**77**(3):W17-W26

[18] Fomel S, Ying L, Song X. Seismic wave extrapolation using low-rank symbol approximation. Geophysical Prospecting. 2013;**61**(3):526-536

[19] Finkelstein B, Kastner R. Finite difference time domain dispersion reduction schemes. Journal of Computational Physics. 2007;**221**:422-438

[20] Finkelstein B, Kastner R. A comprehensive new methodology for formulating FDTD schemes with controlled order of accuracy and dispersion. IEEE Transactions on Antennas and Propagation. 2008;**56**:3516-3525

[21] Etgen J. T. A tutorial on optimizing time domain finite-difference scheme: Beyond Holberg: Stanford Exploration Project Report. 2007;**129**:33-43

[22] Liu Y, Sen MK. A new time-space domain high-order finite difference method for the acoustic wave equation. Journal of Computational Physics. 2009;**228**:8779-8806

[23] Liu Y, Sen MK. Scalar wave equation modeling with time-space domain dispersion-relation-based staggered-grid finite-difference schemes. Bulletin of the Seismological Society of America. 2011;**101**(1):141-159

[24] Zhang JH, Yao ZX. Optimized finite-difference operator for broadband seismic wave modeling. Geophysics. 2013;**78**:A13-A18

[25] Liang WQ, Yang CC, Wang YF, Liu HW. Acoustic wave equation modeling with new time-space domain finite difference operators. Chinese Journal of Geophysics. 2013;**56**(6): 840-850

[26] Ren Z, Liu Y. Acoustic and elastic modeling by optimal time-space-domain staggered-grid finite-difference schemes. Geophysics. 2014;**80**(1):T17-T40

[27] Wang Y, Liang W, Nashed Z, Li X, Liang G, Yang C. Seismic modeling by optimizing regularized staggered-grid finite-difference operators using a time-space-domain dispersion- relationship-preserving method. Geophysics. 2014;**79**(5):T277-T285

[28] Chen H, Zhou H, Zhang Q, Chen Y. Modeling elastic wave propagation using K space operator-based temporal high-order staggered-grid finite-difference method. IEEE Transactions on Geoscience and Remote Sensing. 2017;**55**(2):801-815

[29] Yong P, Huang J, Li Z, Liao W, Qu L, Li Q, Liu P. Optimized equivalent staggered-grid FD method for elastic wave modelling based on plane wave solutions. Geophysical Journal International. 2017;**208**(2):1157-1172

[30] Liu Y, Sen MK. Time-space domain dispersion-relation-based finite-difference method with arbitrary even-order accuracy for the 2D acoustic wave equation. Journal of Computational Physics. 2013;**232**(1):327-345

[31] Liu H, Dai N, Niu F, Wu W. An explicit time evolution method for acoustic wave propagation. Geophysics. 2014;**79**(3):T117-T124

[32] Tan S, Huang L. An efficient finite-difference method with high-order accuracy in both time and space domains for modelling scalar-wave propagation. Geophysical Journal International. 2014a;**197**(2):1250-1267

[33] Tan S, Huang L. A staggered-grid finite-difference scheme optimized in the time-space domain for modeling scalar-wave propagation in geophysical problems. Journal of Computational Physics. 2014b;**276**:613-634

[34] Margrave GF. Numerical Methods of Exploration Seismology with Algorithms in Matlab. Calgary: Department of Geology and Geophysics, the University of Calgary; 2001

[35] Liu Y. Optimal staggered-grid finite-difference schemes based on least-squares for wave equation modelling. Geophysical Journal International. 2014;**197**(2):1033-1047

Chapter 2

Acoustic Analysis of Enclosed Sound Space as well as Its Coupling with Flexible Boundary Structure

Jingtao Du, Yang Liu and Long Liu

Additional information is available at the end of the chapter

http://dx.doi.org/10.5772/intechopen.69967

Abstract

Combustion instability is often encountered in various power systems, a good understanding on the sound field in acoustic cavity as well as its coupling with boundary flexible structure will be of great help for the reliability design of such combustion system. An improved Fourier series method is presented for the acoustic/vibro-acoustic modelling of acoustic cavity as well as the panel-cavity coupling system. The structural-acoustic coupling system is described in a unified pattern using the energy principle. With the aim to construct the admissible functions sufficiently smooth for the enclosed sound space as well as the flexible boundary structure, the boundary-smoothed auxiliary functions are introduced to the standard multi-dimensional Fourier series. All the unknown coefficients and higher order variables are determined in conjunction with Rayleigh-Ritz procedure and differential operation term by term. Numerical examples are then presented to show the correctness and effectiveness of the current model. The model is verified through the comparison with those from analytic solution and other approaches. Based on the model established, the influence of boundary conditions on the acoustic and/or vibro-acoustic characteristics of the structural-acoustic coupling system is addressed and investigated.

Keywords: enclosed sound space, acoustics analysis, structural-acoustic coupling, flexible boundary structure

1. Introduction

Combustion instability is often encountered in various power systems, which will further cause the combustion noise or even the dynamic damage of combustion chamber structure [1]. A good understanding on the vibro-acoustic coupling between the bounded flexible structure and the thermo-driven acoustic oscillation will be of great significance for the correct design of combustion system of various power plants. As an important part of such whole thermos-acoustic coupling system, the acoustics in cavity and its coupling with its flexible boundary also plays an important role. For many years, a lot of research effort has been devoted to the coupled structural-acoustic system.

The acoustic analysis in enclosed space is a classical research topic in acoustics community, and the rectangular cavity is widely used as the theoretical model. Morse and Bolt [2] first introduced the normal modes theory into room acoustics, and developed a non-linear transcendental characteristic equation through combining the assumed sound pressure modes with complex impedance boundary conditions on the walls. Maa [3] derived the transcendental equation for a rectangular room with non-uniform acoustical boundaries which took the same form as that of a uniform acoustic admittance case, while the impedance term was consequently non-uniform on certain wall. Recent studies have been mainly focused on developing more effective root searching algorithms for finding eigensolutions. For instance, Bistafa and Morrissey [4] compared two different numerical procedures: one is the Newton's method and the other is referred to as the homotopic continuation technique based on the numerical integration of differential equations. The roots are searched for the cases from soft walls to the terminal impedance with small increments. They found that the latter procedure is much faster in finding all the possible roots. Naka et al. [5] utilized an interval Newton/generalized bisection (IN/GB) method to find the roots of the non-linear characteristic equation within any given interval for the modal analysis of rectangular room with arbitrary wall impedances.

In many occasions, the acoustic cavity is bounded by the flexible structure, and the interaction between the structural vibration and the acoustic cavity should be taken into account simultaneously for the determination of acoustic field characteristics. Among the existing studies, the most popular modelling approach is the so-called modal coupling theory [6] in which the structural modes *in vacuo* and the acoustic cavity modes with rigid walls need to be determined *a priori*. The two sets of modes are then combined together, via spatial coupling coefficients, to find the response of the coupled system. However, as pointed out by Pan et al. [7, 8], there are two main limitations with the modal coupling theory. One is that such an approach is only suitable for weak coupling and will be inadequate in dealing with strong coupling conditions as in the cases where a very thin plate, a shallow cavity depth or a heavy medium is involved. The other one is related to the use of the rigidly walled cavity modes since then the particle vibrational velocity on contacting surface cannot be determined from the pressure gradient, causing the discontinuity of velocity from the cavity to the vibrational panel. In other words, the basic requirement of velocity continuity on the panel-cavity coupling interface cannot be satisfied by the modal coupling theory. Then, this approach may be problematic when the energy transmission is needed for the analysis, since it will be difficult to calculate the high-order variable using the acoustic mode with rigid wall.

In this chapter, a unified structural-acoustic coupling analysis framework will be introduced for the representative rectangular cavity and its coupled panel structure. The fully coupling system is described in the framework of energy. The Fourier series with supplementary terms is constructed as the admissible functions, which are smoothed in the whole solving domain including the elastic structural and/or impedance acoustic boundary and coupling interface. All the unknown coefficients are solved in conjunction with Rayleigh-Ritz procedure. Since the field functions are sufficiently smooth, the corresponding high-order variables can be calculated straightforwardly.

2. Theoretical formulations

2.1. Acoustic cavity with impedance boundary condition

A rectangular acoustical cavity of dimensions $L_x \times L_y \times L_z$ and the associated coordinate system are sketched in **Figure 1**. In this study, it is assumed that an arbitrary impedance boundary condition is specified on each of the cavity surfaces, that is,

$$\frac{\partial p}{\partial n} = -j\frac{\rho c}{Z_i}kp \tag{1}$$

where $j = \sqrt{-1}$, p is the sound pressure, n denotes the outgoing normal of the surface, ρ and c are respectively the mass density and the sound speed in the acoustic medium, k (= ω/c) is the wavenumber with ω being angular frequency, and Z_i represents the acoustic impedance on the ith surface.

2.2. Improved Fourier series representation of admissible function

It is well known that the modal functions for rigid-walled rectangular cavity are simply the products of cosine functions in three dimensions. Based on the modal superposition principle, the corresponding sound pressure field inside the cavity can be generally expressed as a 3-D Fourier cosine series. However, such a Fourier series representation will become problematic when an impedance boundary condition is specified on one or more of the interior walls. This assertion is evident from Eq. (1) because the left side of the equation is identically equal to zero regardless of the actual value of the right side. This problem is mathematically related to the inability to converge of the traditional Fourier series on the boundaries of a domain under general boundary conditions. To overcome this difficulty, in this study, a 3-D version of an

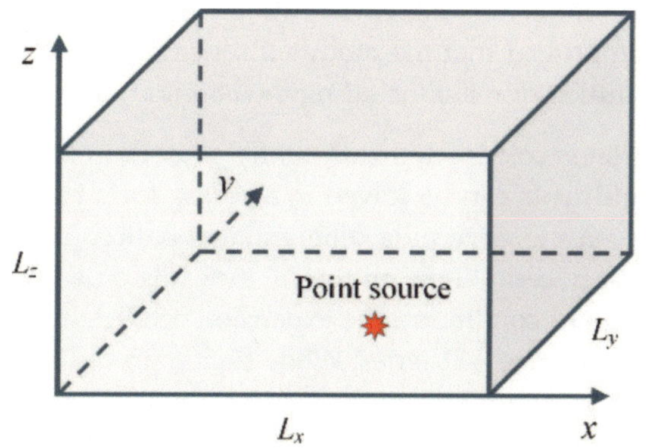

Figure 1. A rectangular cavity with general impedance boundary conditions.

improved Fourier series representation previously developed for the in-plane vibrations of elastically restrained plates will be used to expand the sound pressure inside the cavity [9].

$$
\begin{aligned}
p(x,y,z) = & \sum_{m_x=0}^{\infty}\sum_{m_y=0}^{\infty}\sum_{m_z=0}^{\infty} A_{m_x m_y m_z}\cos\lambda_{m_x}x\cos\lambda_{m_y}y\cos\lambda_{m_z}z \\
& + \sum_{m_x=0}^{\infty}\sum_{m_y=0}^{\infty}\left[\underbrace{\xi_{1L_z}(z)a_{m_x m_y}}_{z=0} + \underbrace{\xi_{2L_z}(z)b_{m_x m_y}}_{z=L_z}\right]\cos\lambda_{m_x}x\cos\lambda_{m_y}y \\
& + \sum_{m_x=0}^{\infty}\sum_{m_z=0}^{\infty}\left[\underbrace{\xi_{1L_y}(y)c_{m_x m_z}}_{y=0} + \underbrace{\xi_{2L_y}(y)d_{m_x m_z}}_{y=L_y}\right]\cos\lambda_{m_x}x\cos\lambda_{m_z}z \\
& + \sum_{m_y=0}^{\infty}\sum_{m_z=0}^{\infty}\left[\underbrace{\xi_{1L_x}(x)e_{m_y m_z}}_{x=0} + \underbrace{\xi_{2L_x}(x)f_{m_y m_z}}_{x=L_x}\right]\cos\lambda_{m_y}y\cos\lambda_{m_z}z
\end{aligned}
\tag{2}
$$

where $\lambda_{ms} = m_s\pi/L_s$, ($s = x$, y or z), and the supplemental functions are defined as

$$\xi_{1L_s}(s) = L_s\zeta_s(\zeta_s - 1)^2,\ \xi_{2L_s}(s) = L_s\zeta_s^2(\zeta_s - 1),\ (\zeta_s = s/L_s) \tag{3}$$

It is easy to verify that

$$\xi_{1L_s}(0) = \xi_{1L_s}(L_s) = \xi'_{1L_s}(L_s) = 0,\ \xi'_{1L_s}(0) = 1 \tag{4}$$

$$\xi_{2L_s}(0) = \xi_{2L_s}(L_s) = \xi'_{2L_s}(0) = 0,\ \xi'_{2L_s}(L_s) = 1 \tag{5}$$

In light of Eqs. (3)–(5), one can understand that the 2-D Fourier series expansions in Eq. (2) mathematically represent the possible non-zero (normal) derivatives of the acoustic pressure on each of the cavity walls, and the 3-D Fourier series a residual pressure field as if the impedance boundary conditions on the cavity walls were modified to being infinite rigid. Mathematically, it can be proved that the modified series solution converges faster and uniformly over the entire solution domain including the boundary walls [10, 11].

Since the pressure solution is constructed sufficiently smooth in the current formulation, the unknown expansion coefficients can be solved in a strong form by letting the series solution simultaneously satisfy both the governing differential equation (Helmholtz equation) inside the cavity and the boundary conditions, on the cavity walls on a point-wise basis. In such a case, because of the boundary conditions, the expansion coefficients for the 2-D series are not fully independent of those for the 3-D series. While such a procedure may be preferred in the context of 'exact' solution, an alternative procedure for obtaining a weak form of solution will be employed here because of its potential benefits in modelling complex acoustic systems consisting of many cavities. The corresponding Lagrangian for the rectangular cavity with arbitrary impedance boundary conditions can be written as

$$L = V - T - W_{ext} \tag{6}$$

where V denotes the total acoustic potential energy stored in the enclosed volume, T represents the total kinetic energy and W_{ext} represents all the work done by the applied sources which include the energy dissipation on the wall surfaces in the current case. These terms can be explicitly expressed as [12].

The total potential energy V is

$$V = \frac{1}{2\rho_0 c_0^2} \int_V p^2 dV = \frac{1}{2\rho_0 c_0^2} \int_0^{L_x} \int_0^{L_y} \int_0^{L_z} p^2(x, y, z) dxdydz \tag{7}$$

where c_0 is the speed of sound, and ρ_0 is the mass density of the medium in the cavity.

The total kinetic energy T is given as

$$\begin{aligned} T &= \frac{1}{2\rho_0 \omega^2} \int_V (\text{grad } p)^2 dV \\ &= \frac{1}{2\rho_0 \omega^2} \int_V \left[\left(\frac{\partial p}{\partial x}\right)^2 + \left(\frac{\partial p}{\partial y}\right)^2 + \left(\frac{\partial p}{\partial z}\right)^2 \right] dV \\ &= \frac{1}{2\rho_0 \omega^2} \int_0^{L_x} \int_0^{L_y} \int_0^{L_z} \left[\left(\frac{\partial p}{\partial x}\right)^2 + \left(\frac{\partial p}{\partial y}\right)^2 + \left(\frac{\partial p}{\partial z}\right)^2 \right] dxdydz \end{aligned} \tag{8}$$

where grad p means the gradient of sound pressure.

By using the relationship between the sound pressure and the particle velocity on impedance surface, the dissipated acoustic energy can be calculated from

$$W_{wall} = -\frac{1}{2} \int_S \frac{p^2}{j\omega} \frac{1}{Z} dS = -\frac{1}{2} \int_S \frac{p^2}{j\omega} Y dS \tag{9}$$

where Z is the complex acoustic impedance of the wall surface, and Y is the complex acoustic admittance which is defined as the inverse of impedance, namely, $Y = 1/Z$. For the non-uniform distributed on a wall surface to account for practical complications, for example, the acoustic admittance on wall $z = 0$ can be generally described as $Y_{z0}(x, y) = Y_A \times h_{z0}(x, y)$ where Y_A is the complex amplitude and $h_{z0}(x, y)$ is a strength distribution function. In this study, to unify the formulations and simplify the subsequent calculations, any specified admittance distribution, such as $Y_{z0}(x, y)$, will be expanded into double Fourier series as

$$Y_{x0}(y, z) = \sum_{n_y=0}^{\infty} \sum_{n_z=0}^{\infty} \tilde{Y}_{x0}^{n_y n_z} \cos \lambda_{n_y} y \cos \lambda_{n_z} z \tag{10}$$

where $\lambda_{ns} = n_s \pi / L_s$, ($s = y$ or z). In actual numerical calculations, all such Fourier series expansions will be truncated to $n_y = N_y$ and $n_y = N_z$. The non-uniform impedance distributions on other wall surfaces can be treated in the same way.

The work done by a sound source inside cavity can be represented as

$$W_S = -\frac{1}{2}\int_V \frac{pQ}{j\omega}dv \tag{11}$$

where Q is the distribution function of a sound source. For a point source located at (x_e, y_e, z_e) inside the cavity, we have

$$Q = Q_0\delta(x - x_e)\delta(y - y_e)\delta(z - z_e) \tag{12}$$

where Q_0 is the volume velocity amplitude of the sound source, and δ is Dirac delta function.

Substituting Eqs. (7)–(12) into the acoustic Lagrangian, Eq. (6), and applying the Rayleigh-Ritz procedure against each of the unknown Fourier series coefficient, a system of linear algebra equations can be derived as

$$(\mathbf{K} + \omega\mathbf{Z} + \omega^2\mathbf{M})\mathbf{E} = \mathbf{Q} \tag{13}$$

where **K** and **M** are the stiffness and mass matrixes of the acoustic system, respectively; **Z** is the damping matrix due to the dissipative effect of the impedance boundary conditions over the cavity walls and **Q** is the external load vector.

In order to determine the modal characteristics of the acoustic cavity, one needs to solve the characteristic equation by setting the external load vector **Q** (on the right side of Eq. (14)) to zero. Since the resulting equation will involve the first-order and second-order terms of oscillation frequency, it is usually rewritten in state space form [13]

$$(\mathbf{R} - \omega\mathbf{S})\mathbf{G} = \mathbf{0} \tag{14}$$

where $\mathbf{R} = -\begin{bmatrix} [0] & \mathbf{K} \\ \mathbf{K} & \mathbf{Z} \end{bmatrix}$, $\mathbf{S} = \begin{bmatrix} -\mathbf{K} & [0] \\ [0] & \mathbf{M} \end{bmatrix}$, $\mathbf{G} = \begin{Bmatrix} \mathbf{E} \\ \mathbf{F} \end{Bmatrix}$ and $\mathbf{F} = \omega\mathbf{E}$.

2.3. Vibro-acoustic coupling of panel-cavity system

The above formulation is mainly about the modelling of pure acoustic cavity, in many situations, the cavity is bounded by the flexible structure, such as the combustion chamber. For such structural-acoustic coupling system, the vibration of flexible boundary structure and the acoustic filed will couple together. As a classical example, the rectangular panel-cavity is often used as the analysis example for the structural-acoustic coupling study.

As shown in **Figure 2**, an elastically restrained plate is one of the surfaces enclosing a rectangular acoustical cavity (other five surfaces are assumed to be perfectly rigid for simplicity). Suppose that the plate is excited by a normal concentrated force *F*. The vibration of the plate will cause sound waves radiated into the cavity, and the cavity will in turn affect the panel vibration by applying sound pressure to the fluid-structure interface. While the phenomenon is described as causal event, it actually defines a coupled structure-acoustical system in which the two different physical processes affect each other and have to be determined simultaneously by considering the coupling conditions at the interface.

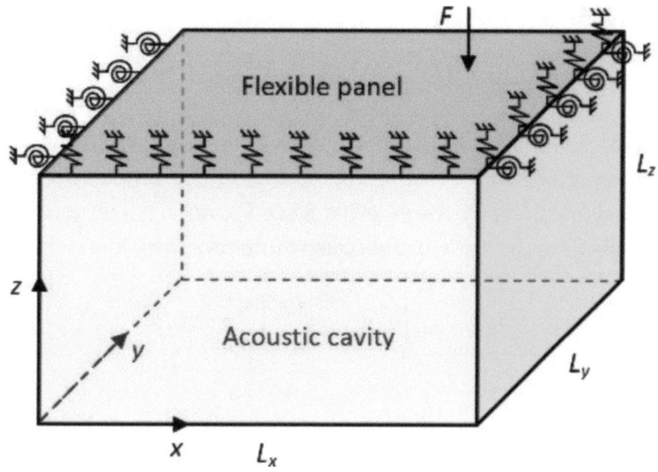

Figure 2. A rectangular acoustic cavity bounded by a flexible panel with general boundary conditions.

Although such structural-acoustic coupling system can be analysed by solving the governing equation and boundary conditions, simultaneously. Similar to the above acoustic analysis of enclosed sound space, the energy principle can also give the sufficiently accurate prediction of vibro-acoustic behaviour, when the admissible functions are constructed smooth enough. For the transverse vibration of a rectangular plate with general elastic boundary supports, its displacement function will be invariantly expanded into an improved Fourier series as [14].

$$w(x,y) = \sum_{m=0}^{\infty}\sum_{n=0}^{\infty} A^p_{mn}\cos\lambda_{L_x m}x\cos\lambda_{L_y n}y + \sum_{l=1}^{4}\left(\zeta_{lL_y}(y)\sum_{m=0}^{\infty}c^l_m\cos\lambda_{L_x m}x + \zeta_{lL_x}(x)\sum_{n=0}^{\infty}d^l_n\cos\lambda_{L_y n}y\right) \tag{15}$$

where $\lambda_{Lxm} = m\pi/L_x$, $\lambda_{Lyn} = n\pi/L_y$ and the superscript p with A_{mn} means the Fourier coefficients for the panel displacement. The supplementary functions $\zeta_{lLx}(x)$ and $\zeta_{lLy}(y)$ are introduced to account for all the possible discontinuities with the first and third partial derivatives (with respect to x or y) of the displacement function along each edge of the plate.

For the panel cavity considered here, the main attention will be paid to the structural-acoustic continuity, with the other walls kept as rigid. The acoustic pressure filed function is constructed as [15]

$$\begin{aligned} p(x,y,z) = &\sum_{m_x=0}^{\infty}\sum_{m_y=0}^{\infty}\sum_{m_z=0}^{\infty} A^a_{m_x m_y m_z}\cos\lambda_{m_x}x\cos\lambda_{m_y}y\cos\lambda_{m_z}z \\ &+ \xi_{2L_z}(z)\sum_{m_x=0}^{\infty}\sum_{m_y=0}^{\infty} b^a_{m_x m_y}\cos\lambda_{m_x}x\cos\lambda_{m_y}y \end{aligned} \tag{16}$$

where m_x, m_y and m_z are all integers, describing the spatial characteristic of a particular mode, $Am_xm_ym_z$ is the complex modal amplitude corresponding to the (m_x, m_y, m_z) mode, and $\lambda_{ms} = m_s\pi/L_s$ $(s = x, y \text{ or } z)$.

The Lagrangian for the plate structure can be expressed as

$$L_{panel} = U_{panel} - T_{panel} - W_{panel} + W_{a\&p} \tag{17}$$

where U_{panel} is the total potential energy associated with the transverse deformation of the panel and the potential energy stored in the restraining springs; T_{panel} denotes the total kinetic energy of the plate; W_{panel} is the work done by the external force F; and $W_{a\&p}$ represents the work done by the sound pressure acting on the structural-acoustic interface which is calculated from

$$W_{a\&p} = \int_S wpdS = \int_0^{L_x}\int_0^{L_y} wpdxdy \tag{18}$$

The total potential and kinetic energy for the elastic plate can be explicitly expressed as

$$\begin{aligned} U_{panel} =& \frac{D}{2}\int_0^{L_x}\int_0^{L_y}\left\{\left(\frac{\partial^2 w}{\partial x^2}\right)^2 + \left(\frac{\partial^2 w}{\partial y^2}\right)^2 + 2\mu\frac{\partial^2 w}{\partial x^2}\frac{\partial^2 w}{\partial y^2} + 2(1-\mu)\left(\frac{\partial^2 w}{\partial x \partial y}\right)^2\right\}dxdy \\ &+ \frac{1}{2}\int_0^{L_y}\left[k_{x0}w^2 + K_{x0}\left(\frac{\partial w}{\partial x}\right)^2\right]_{x=0} dy + \frac{1}{2}\int_0^{L_y}\left[k_{xL_x}w^2 + K_{xL_x}\left(\frac{\partial w}{\partial x}\right)^2\right]_{x=L_x} dy \\ &+ \frac{1}{2}\int_0^{L_x}\left[k_{y0}w^2 + K_{y0}\left(\frac{\partial w}{\partial y}\right)^2\right]_{y=0} dx + \frac{1}{2}\int_0^{L_y}\left[k_{yL_y}w^2 + K_{yL_y}\left(\frac{\partial w}{\partial y}\right)^2\right]_{y=L_y} dx \end{aligned} \tag{19}$$

and

$$T_{panel} = \frac{1}{2}\int_0^{L_x}\int_0^{L_y} \rho h\left(\frac{\partial w}{\partial t}\right)^2 dxdy = \frac{1}{2}\rho h\omega^2\int_0^{L_x}\int_0^{L_y} w^2 dxdy \tag{20}$$

where ρ and h are the mass density and thickness of the plate structure, respectively.

The Lagrangian for the acoustic cavity is

$$L_{cavity} = U_{cavity} - T_{cavity} - W_{p\&a} \tag{21}$$

where U_{cavity} is the total potential energy for the acoustic cavity, T_{cavity} is the total kinetic energy of the particle vibrations inside the cavity and $W_{p\&a}$ denotes the work due to the panel vibration. The (pressure and velocity) continuity conditions the solid-fluid interface implies a reciprocity relationship, that is, $W_{p\&a} = W_{a\&p}$.

Substituting Eqs. (15) and (16) into Eqs. (17) and (21) and minimizing them against the unknown Fourier coefficients, one is able to obtain the final system in matrix form as

$$\left\{\begin{bmatrix} \mathbf{K_p} & \mathbf{C_{a\&p}} \\ \mathbf{0} & \mathbf{K_a} \end{bmatrix} - \omega^2\begin{bmatrix} \mathbf{M_p} & \mathbf{0} \\ -\mathbf{C}^T_{\mathbf{a\&p}} & \mathbf{M_a} \end{bmatrix}\right\}\begin{bmatrix} \mathbf{W} \\ \mathbf{P} \end{bmatrix} = \begin{Bmatrix} \mathbf{F_p} \\ \mathbf{0} \end{Bmatrix} \tag{22}$$

Once the Fourier coefficient vectors **W** and **P** are solved from Eq. (22), the structural displacement on the plate and the sound pressure in the cavity can be readily determined by using Eqs. (15) and (18). If one is only interested in the modal parameters of the coupled structural-acoustic system, they can be simply obtained from solving a standard matrix characteristic equation by setting to zero the loading vector on right hand side of Eq. (22), instead of searching the singularities (the poles) of the modal coefficients or extracting the resonant peaks from the frequency response functions of the coupled system. It should be noted that although only one of cavity surfaces is considered movable in the above discussion, the present method can be readily extended to an acoustic cavity bounded by multiple plate structures.

3. Numerical examples and discussion

In this section, numerical examples will be presented to demonstrate the effectiveness and reliability of the proposed method, then based on model established, the vibro-acoustic behaviour of the cavity as well as its coupling system with the flexible panel will be studied. The first example involves a rectangular cavity with each of its walls being perfectly rigid. The related parameters are as follows: the dimensions are $L_x \times L_y \times L_z = 0.7 \times 0.6 \times 0.5$ m^3, the air density is ρ_{air} = 1.21 kg/m^3 and the speed of sound is c_0.= 340 m/s. For a non-dissipative wall, its acoustic impedance is described only by an imaginary part.

Table 1 shows a comparison of the first six natural frequencies using the familiar analytic solution and from the current method. It should be mentioned that the current results were calculated by truncating the series expansions in Eq. (2) to $M_x = M_y = M_z = 3$. The 'perfect' match between the current results and the exact solution partially indicates the excellent mathematical characteristic of the proposed series solution in terms of its convergence and accuracy. Although only the first six modes were compared in **Table 1**, a nice agreement for other higher order modes is also evident from the acoustic response curves presented in **Figure 3**. In the subsequent calculations, all the Fourier series expansions will be truncated to $M = 3$ in each direction unless otherwise specified.

Now, place a point source of strength $Q_0 = 2 \times 10^{-5}$ m^3/s into the acoustic cavity at position (L_x/10, L_y/10 and L_z/10). To account for the air absorption, a modal damping ratio $\eta = 0.01$ is assumed for each acoustic mode. For a relatively small acoustic damping ratio, the dissipative effect can be accounted for simply through introducing a complex wavenumber $k' = k(1 - j\xi)$ [16]. In this

	Natural frequency (Hz)					
	1	2	3	4	5	6
Current	0.00	242.86	283.33	340.00	373.17	417.83
Analytical	0.00	242.86	283.33	340.00	373.17	417.83

Table 1. The six lowest natural frequencies for a cavity with perfectly rigid walls.

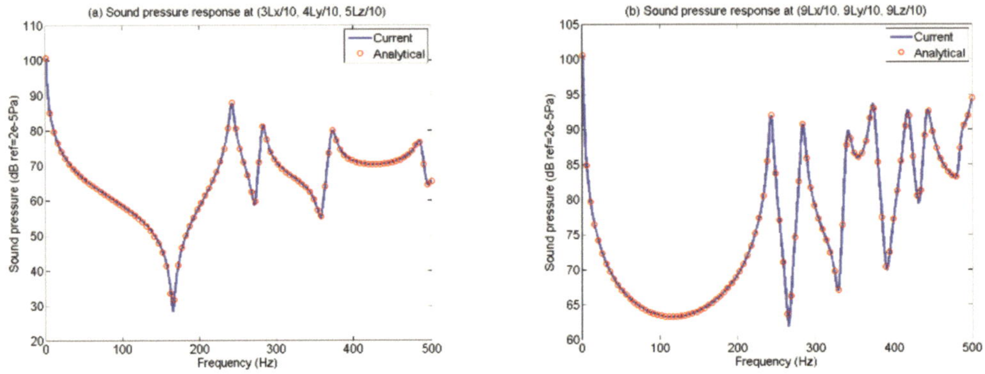

(a) (b)

Figure 3. Sound pressure responses inside the cavity at: (a) ($3L_x/10$, $4L_y/10$, $5L_z/10$) and (b) ($9L_x/10$, $9L_y/10$, $9L_z/10$).

method, a complex sound speed $\tilde{c} = c(1 - j\eta)$ is used instead. Since $k' = \omega/\tilde{c}$ (or, $\omega/[c(1 - j\eta)] = k(1 - j\xi)$), it is easy to see that $\eta = -\xi/(1 - j\xi)$. The sound pressure levels at two observation points, ($3L_x/10$, $4L_y/10$, $5L_z/10$) and ($9L_x/10$, $9L_y/10$, $9L_z/10$), are plotted in **Figure 3** in the frequency range of 0–500 Hz. For comparison, the results obtained based on the superposition of 245 analytical modes are also presented. An excellent comparison between the two predictions is seen over the entire frequency range [16].

Another extreme case of the non-dissipative boundary conditions is the so-called pressure release (or zero-pressure), which is described by infinitely small pure imaginary impedance on surface ($j10^{-5}$ in the actual calculations). Suppose that a cavity of $2.12 \times 6.06 \times 2.12\ m^3$ has a pressure-release condition at surface $y = L_y$ (= 6.06 m), and the rest walls are perfectly rigid. A point source with strength as $Q_0 = 2 \times 10^{-5}\ m^3/s$ is placed at (1.86, 0.26 and 0.26 m). To account for the larger dimension in the y direction, more expansion terms are retained accordingly, that is, $M_y = 7$ as compared with $M_x = M_z = 3$.

The sound pressure at (0.1, 5.96 and 2.02 m) is plotted in **Figure 4** as a function of frequency. This problem was previously solved by using an equivalent source technique, and the result was also shown in **Figure 4** as a reference. It is seen that the two predictions are in a good agreement. However, slight separation between them can be noticed as frequency increases. This is probably caused by the possible loss of the accuracy of the equivalent source technique due to the use of an insufficient number of equivalent sources at higher frequencies. Plotted in **Figure 5** are sound pressure fields inside the cavity at 14 and 42 Hz, respectively. Since two frequencies are very close to the first two resonance frequencies (refer to **Figure 4**), the distributions essentially resemble the first and second acoustic mode shapes. It is observed from **Figure 5** that the sound pressure decreases rapidly in approaching to (and eventually vanishes on) the wall of $y = L_y$. The pressure fields are basically uniform on the cross-section perpendicular to y-axis. These two pressure patterns may be considered to evolve the familiar (1, 0, 0) and (2, 0, 0) modes for the cavity with each wall being perfectly rigid. The existence of the zero-pressure wall at $y = L_y$ causes the nodal surfaces to shift towards it.

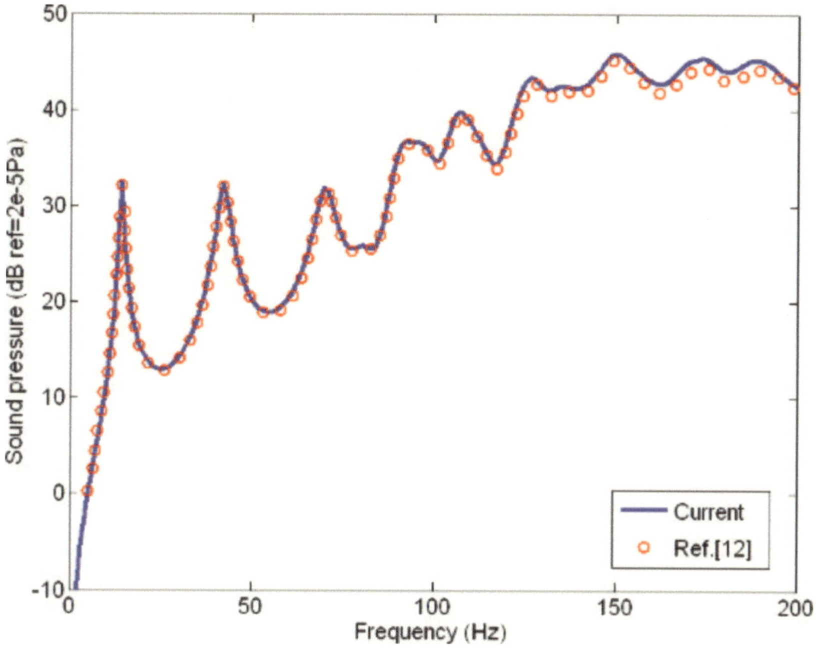

Figure 4. Sound pressure at (0.1, 5.96 and 2.02 m) inside the cavity with pressure release boundary specified on the wall $y = L_y$.

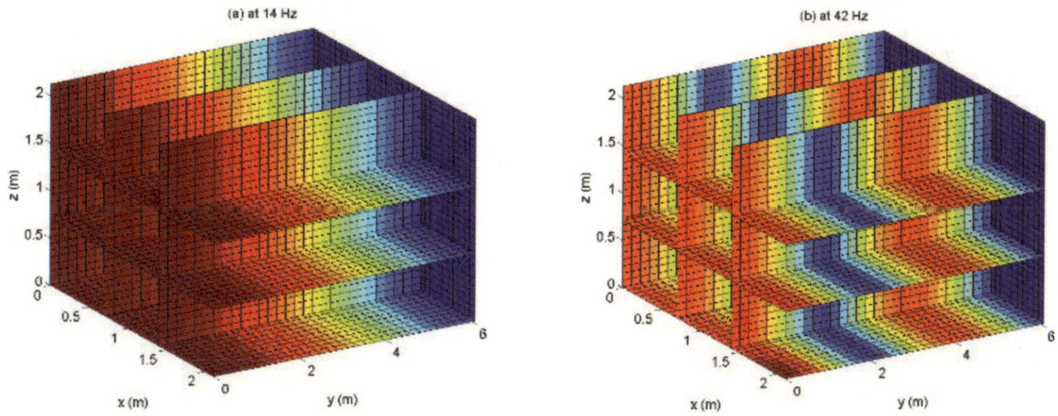

Figure 5. Sound pressure fields with a pressure release described by an infinitely small impedance on the surface ($j10^{-5}$ in the actual calculations) on surface $y = L_y$ (=6.06m) at: (a) 14 Hz and (b) 42 Hz.

To better understand the effect of impedance boundary conditions on the acoustical characteristic of an enclosed space, the frequency responses at the observing point (0.1, 5.96 and 2.02 m) are plotted in **Figure 6** for a wide range of impedance values from $j10^{-5}$ to $j10^{5}$ specified on surface $y = L_y$ (= 6.06 m), while the other walls are kept acoustically rigid. For small impedances, $Z_i \le j10^4$, both the resonance frequencies and response amplitudes show a strong

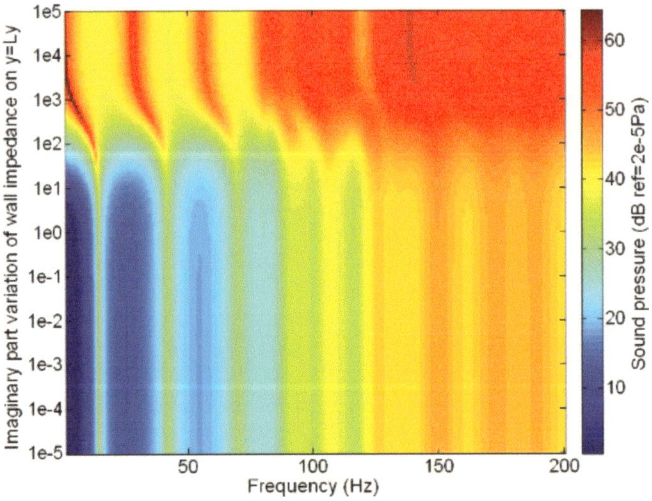

Figure 6. Effect of varying impedance boundary condition on the sound pressure response at observing position (0.1, 5.96 and 2.02 m), in which the impedance variation is specified on surface $y = L_y$ (=6.06 m) while the other walls are kept acoustically rigid.

dependence on the specified impedance value. However, when the wall becomes sufficiently rigid, $Z_i > j10^4$, a further increase of the impedance will have little effect on the pressure field.

In the aforementioned examples, the results are mainly on the acoustic cavity analysis, as demonstrated in the above formulations, the current modelling framework can be used for the treatment of vibro-acoustics analysis of panel-cavity system by simply including the vibrational energy in the whole description. The model will be first valeted, and then, the main emphasis will be put on the model validation and the influence of structural boundary condition on the coupling characteristics of such panel-cavity system.

For the model verification on the modal parameter prediction, consider a problem previously studied in Ref. [17] where an acoustic cavity ($L_x \times L_y \times L_z = 0.2032$ m $\times$ 0.4064 m $\times$ 0.6096 m) is coupled to a simply supported rectangular plate ($L_x \times L_y = 0.2032$ m $\times$ 0.4064 m) of thickness 1.524 mm. The other five walls of the cavity are considered as acoustically rigid. The material properties of the plate are specified as: Poisson's ratio $\mu = 0.3$, Young's modulus $E = 71 \times 10^9$ Pa and mass density $\rho_{panel} = 2700$ kg/m^3. The density of and sound speed in the air cavity are $\rho_{air} = 1.21$ kg/m^3 and $c_0 = 344$ m/s, respectively. In the current solution method, the simply supported boundary condition can be easily realized by respectively setting the stiffnesses of the rotational and translational springs to zero and infinity which is actually represented by a very large number, 5×10^9, in the numerical calculations.

Table 2 shows the first 20 natural frequencies of the coupled panel-cavity system. The data from Ref. [17] were also listed there for comparison. A nice agreement can be observed between these two sets of results with the largest difference being less than 0.35%. In this example, the Fourier series is truncated to $M_p = N_p = 12$ for plate displacement and to $M_{xa} = M_{ya} = M_{za} = 3$ for the cavity pressure.

Mode order	Ref. [17] (Hz)	Current (Hz)	Difference (%)
1	113.91	114.06	0.13
2	177.48	178.04	0.32
3	280.71	281.02	0.11
4	295.97	296.62	0.22
5	379.77	379.71	0.02
6	423.05	423.11	0.01
7	447.32	447.96	0.14
8	448.06	449.21	0.26
9	511.5	511.52	0.00
10	559.9	561.49	0.28
11	565.89	565.9	0.00
12	650.43	652.6	0.33
13	706.59	706.6	0.00
14	717.41	719.91	0.35
15	829.09	826.97	0.26
16	846.99	847	0.00
17	847.03	847.04	0.00
18	850.95	850.59	0.04
19	892.84	892.92	0.01
20	893.4	893.39	0.00

Table 2. The first 20 modal frequencies for the coupled panel-cavity system.

Next, we will examine the effect of cavity depth on the modal parameters. As in Ref. [18], the cavity dimensions are chosen as: $L_x \times L_y \times L_z = 0.2\ \text{m} \times 0.2\ \text{m} \times H_{depth}$. The cavity is enclosed at $z = L_z$ by a simply supported copper panel, and its remaining five walls are assumed to be perfectly rigid. The panel thickness is 0.9144 mm, and its material properties are mass density $\rho_s = 8440\ \text{kg/m}^3$, Poisson's ratio $\mu = 0.35$ and Young's modulus $E = 105 \times 10^9$ Pa. The acoustic medium is air having the same properties as in the first example. **Table 2** gives the first six modal frequencies of the coupled system with a good comparison with the results previously presented in Ref. [18].

From the standpoint of structural vibration, the acoustic cavity may be approximately viewed as Winkler springs with a probably non-uniform stiffness distribution over the area of the panel, depending upon frequency. To understand its significance, the effects of varying edge restraining stiffnesses on the fundamental frequency of the coupled system are studied for a range of cavity depths. In this analysis, the copper panel is assumed to be uniformly supported along all four edges. Shown in **Figure 7(a)** are the results for a configuration in which, by keeping the rotational stiffness to zero, the stiffness for the translational spring is increased from zero (completely free) to infinity (simply supported). It is evident that reducing the cavity

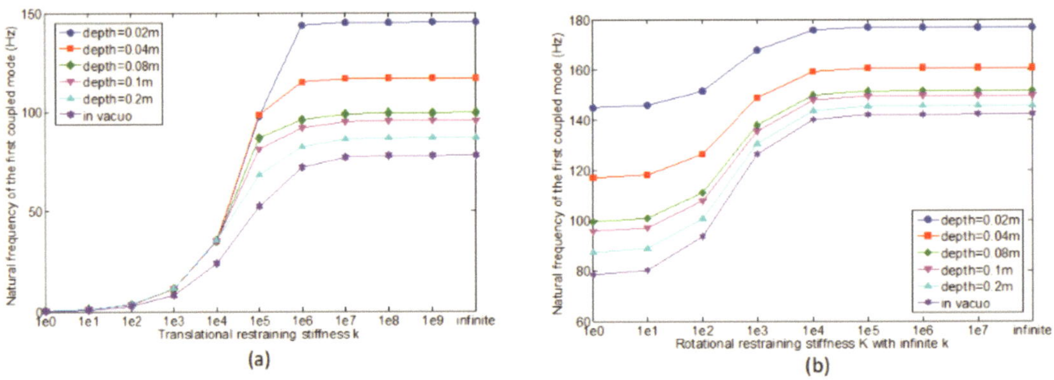

Figure 7. Effect of the boundary restraining stiffness on the first modal frequency of the coupled system: (a) varying the stiffness of the translational springs; (b) varying the rotational stiffness.

depth is equivalent to increasing the stiffness of the Winker springs. For small restraining stiffness, $k \leq 10^3$, the first mode is manifested in a piston-like motion, and the fundamental frequency is primarily determined by the edge restraints. However, as the edge springs become sufficiently strong, $k \geq 10^7$, the air cavity starts to take over as a dominating factor in affecting the value of the fundamental frequency. For intermediate values, $10^4 \leq k \leq 10^6$, the fundamental frequency tends to show a strong dependence on the stiffness of the edge restraints.

By letting the translational springs be infinitely rigid, we now add rotational restraints to the edges. The fundamental frequency curves are plotted in **Figure 7(b)** for various combinations of cavity depths and spring stiffnesses. Again, it is seen that the decreasing the cavity depth is equivalent to increasing the stiffness of the Winkler springs and hence the fundamental frequency of the panel wall. The above results also indicate that there tends to exist stronger structural acoustic coupling for a thinner air gap. This statement may have a meaningful implication to the design of double-walled sound isolation.

We will now direct our attention to the vibro-acoustic responses of the coupled system. For validation, the cavity dimensions will be modified to: $L_x \times L_y \times L_z = 1.5\ \text{m} \times 0.3\ \text{m} \times 0.4\ \text{m}$. The top surface of the cavity is a simply supported plate of thickness h = 5 mm. Other relevant material properties are given as: $\rho_{\text{Al}} = 2770\ \text{kg/m}^3$, Young's modulus $E = 71 \times 10^9$ Pa, Poisson's ratio $\mu = 0.33$, air density $\rho_{\text{air}} = 1.21\ \text{kg/m}^3$ and speed of sound $v = 340$ m/s. The damping ratio $\xi = 0.01$ is used for both the plate and air cavities. A unit force is applied to the plate at point $(13L_x/30, L_y/2)$. This model was studied before by Kim and Brennan [19] using an impedance and mobility approach which is essentially the same as the modal coupling theory. **Figure 8** shows the velocity responses at two separate locations, $(13L_x/30, L_y/2)$ and $(16L_x/30, L_y/3)$. Plotted in **Figure 9** are the sound pressures inside the cavity at $(4L_x/10, L_y/2, L_z/2)$ and $(L_x/2, L_y/2, L_z/2)$. The reference values in the dB scales are 10^{-9} m/s for velocity and 2×10^{-5} Pa for sound pressure. It is seen that both the resonant peaks (namely, natural frequencies of the coupled system) and magnitude of the calculated vibrational and acoustic responses match very well with the analytical predictions from the impedance and mobility approach [19],

http://dx.doi.org/10.5772/intechopen.69967

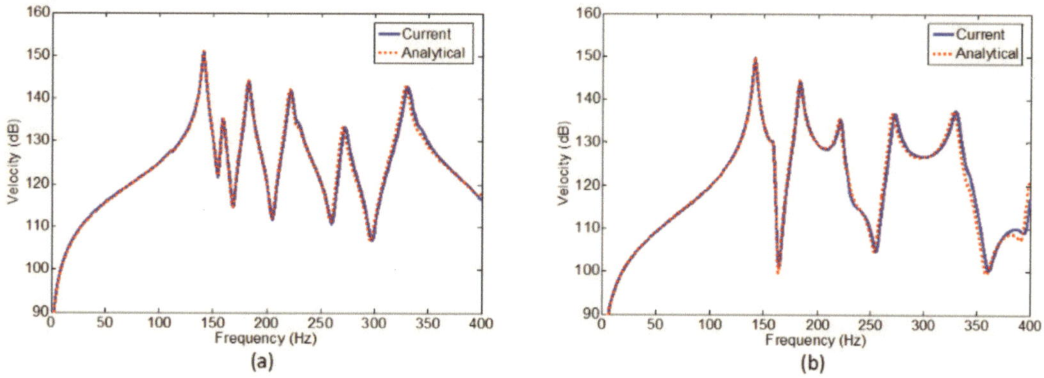

Figure 8. Velocity responses of the plate at: (a) ($13L_x/30$, $L_y/2$) and (b) ($16L_x/30$, $L_y/3$).

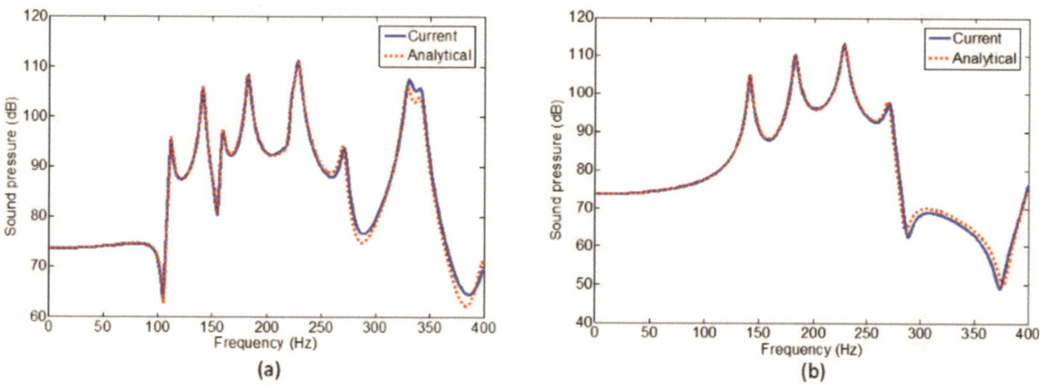

Figure 9. Sound pressure responses at: (a) ($4L_x/10$, $L_y/2$, $L_z/2$) and (b) ($L_x/2$, $L_y/2$, $L_z/2$).

which are plotted as the dotted curves in **Figures 8** and **9**. In this analysis, the Fourier series is truncated to $M_p = N_p = 12$ for the plate displacement and to $M_{xa} = 5$, $M_{ya} = M_{za} = 3$ for the cavity pressure.

As mentioned earlier, the conventional modal coupling theory suffers a velocity discontinuity problem at the fluid-structure interface, that is, the particle velocity on/near the interface cannot correctly calculated from the pressure gradient. However, this velocity continuity requirement is faithfully enforced in the current method. To illustrate this point, **Figure 10(a)** shows the velocity response at ($3L_x/10$, $L_y/4$, L_z) on the panel. Because of the relatively large length-to-width ratio ($L_x/L_y = 1.5/0.3 = 5$), it is reasonable to include more x-related terms in the Fourier series for the cavity pressure to better capture the faster variation of pressure gradient in the x-direction. It is seen that setting the truncation number to M_x = 10 has effectively ensured the velocity continuity on the interface. The direct and indirect velocities are also compared in **Figure 10(b)** at a different point, ($L_x/4$, $L_y/4$, L_z), on the interface. Plotted in **Figure 11** are the comparisons of the velocity responses at ($3L_x/10$, $4L_y/5$, z) predicted by the

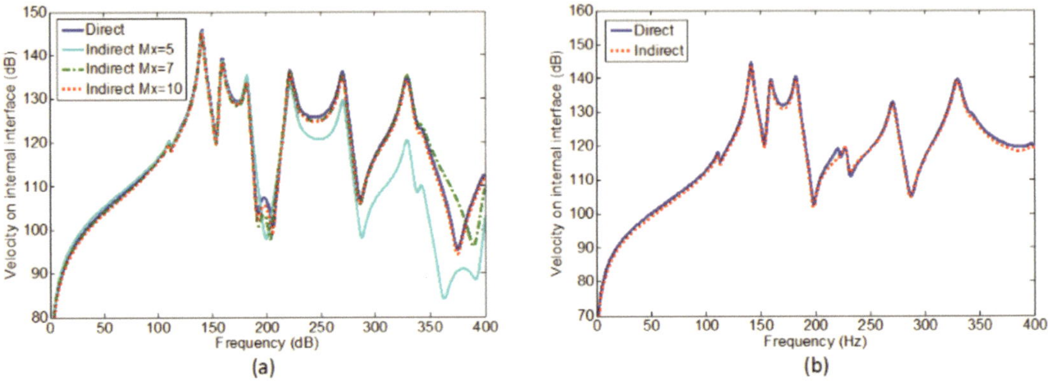

Figure 10. Velocity responses on the interface at: (a) $(3L_x/10, L_y/4)$ and (b) $(L_x/4, L_y/4)$.

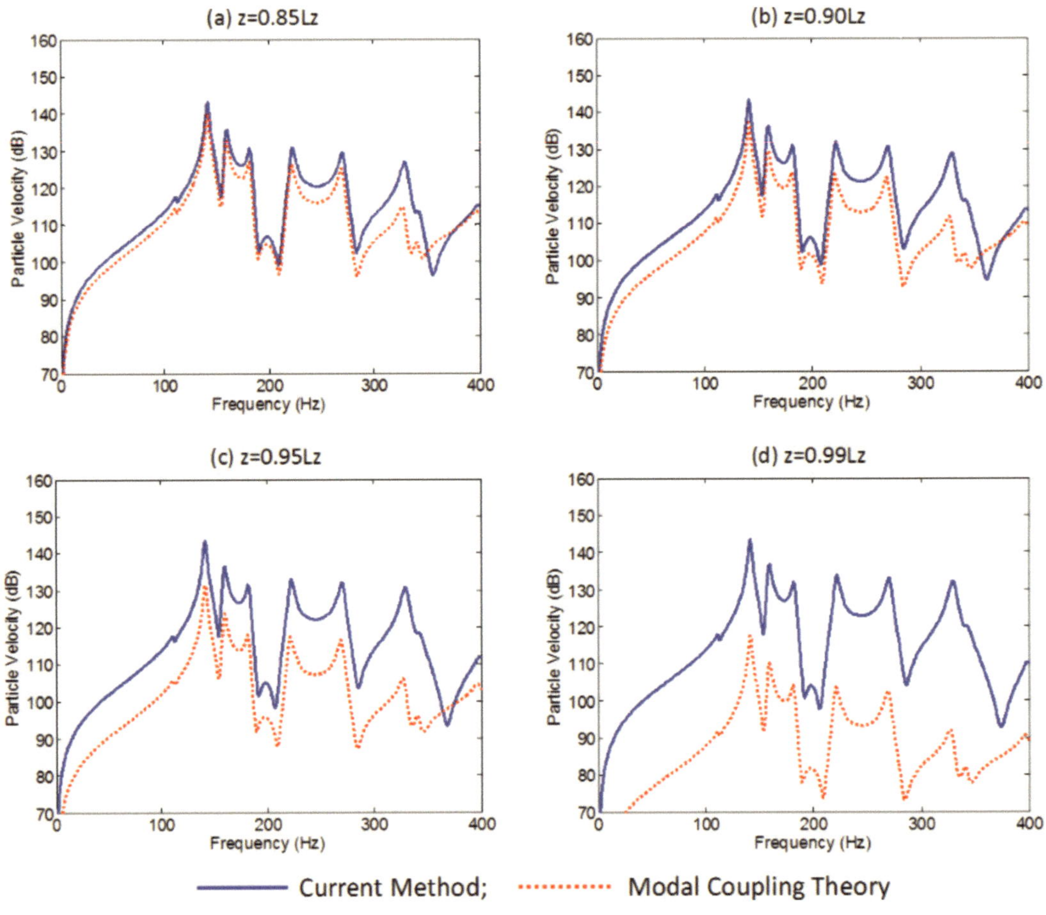

Figure 11. Comparison of the current method and modal coupling theory on predicting the particle velocity response at $(3L_x/10, 4L_y/5, z)$ with different spatial coordinates along z-axis: (a) $z = 0.85L_z$; (b) $z = 0.90L_z$; (c) $z = 0.95L_z$; and (d) $z = 0.99L_z$.

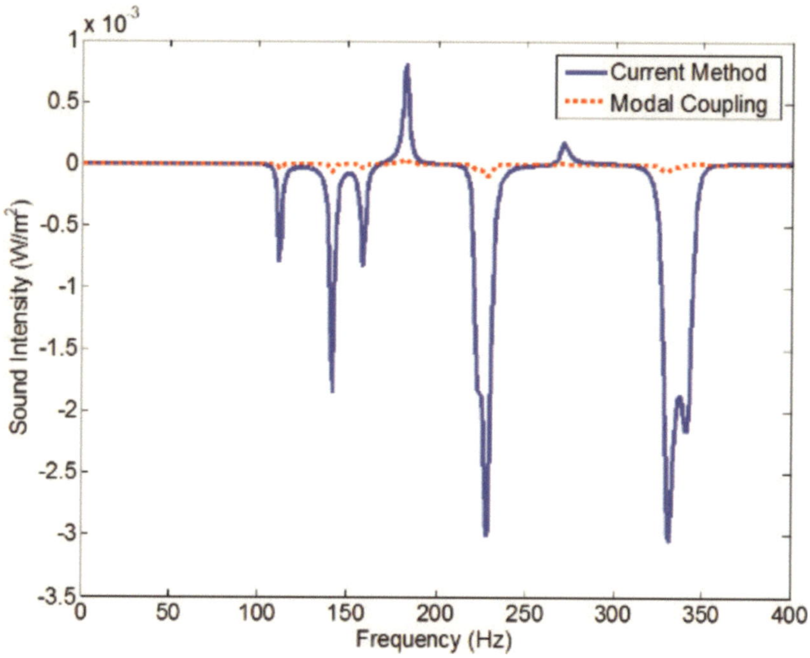

Figure 12. Sound intensity at ($3L_x/10$, $4L_y/5$, $0.99L_z$) predicted by current method and modal coupling theory.

current method and modal coupling theory for different z-values: $z = 0.85L_z$, $0.90L_z$, $0.95L_z$ and $0.99L_z$. It is observed that the particle velocity can be better estimated by using the current method, particularly for the cases very close to the panel-cavity interface. Once the reliable prediction has been made for the particle velocity, other variables of interest such as sound intensity can be calculated. **Figure 12** presents the calculated sound intensity near the interface at ($3L_x/10$, $4L_y/5$, $0.99L_z$). Since sound intensity is a vector, the negative value indicates that its direction is opposite to the z-axis. It can be found that the acoustic energy is not always transmitted from the vibrating panel into acoustical cavity in the whole frequency range at this observing point. The calculated results using modal coupling theory are also shown there; obviously, such an approach cannot be correctly used to predict sound intensity due to its poor accuracy with predicting particle velocity from pressure gradient in the vicinity of the vibrating boundary structure.

4. Conclusions

An improved Fourier series method is presented for the acoustic/vibro-acoustic modelling of acoustic cavity as well as its coupling with flexible boundary structure. The coupled system is described in a unified pattern by using the energy description. With the aim to construct the structural-acoustic admissible functions smooth sufficiently in the whole solving domain, boundary-smoothed auxiliary functions are introduced to the standard multi-dimensional Fourier series on the system boundary as well as the coupling interface. In conjunction with

Rayleigh-Ritz procedure, all the unknown coefficients can be easily derived, and the relevant higher order acoustic variables, such as energy power flow, can be determined straightforwardly.

The theoretical formulation is implemented in the Matlab environment. Numerical results are presented to illustrate the effectiveness and efficiency of the proposed model. The correctness and reliability is then verified by comparing with those from other method or numerical solution. Based on the model established, influence of boundary condition on the acoustic or structural-acoustic coupling characteristics is addressed and investigated in details. This work can present an efficient analysis tool for the acoustic or structural-acoustic analysis of the enclosed sound space and flexible structure. This work shows that the desired modal characteristics of coupling system can be obtained by adjusting boundary conditions properly.

Acknowledgements

This work was supported by the Fundamental Research Funds for the Central Universities (HEUCFP201758).

Author details

Jingtao Du*, Yang Liu and Long Liu

*Address all correspondence to: dujingtao@hrbeu.edu.cn

College of Power and Energy Engineering, Harbin Engineering University, Harbin, PR China

References

[1] Kelsall G, Troger C. Prediction and control of combustion instabilities in industrial gas turbines. Applied Thermal Engineering. 2004;**24**:1571-1582. DOI: 10.1016/j.applthermaleng.2003.10.025

[2] Morse PM, Bolt RH. Sound waves in rooms. Reviews of Modern Physics. 1944;**16**:69-150. DOI: 10.1103/RevModPhys.16.69

[3] Maa DY. Non-uniform acoustical boundaries in rectangular rooms. Journal of the Acoustical Society of America. 1940;**12**:39-52. DOI: 10.1121/1.1916070

[4] Bistafa SR, Morrissey JW. Numerical solutions of the acoustic eigenvalue equation in the rectangular room with arbitrary (uniform) wall impedances. Journal of Sound and Vibration. 2003;**263**:205-218. DOI: 10.1016/S0022-460X(02)01123-9

[5] Naka Y, Oberai AA, Shinn-Cunningham BG. Acoustic eigenvalues of rectangular rooms with arbitrary wall impedances using the interval Newton/generalized bisection method. Journal of the Acoustical Society of America. 2005;**118**:3662-3671. DOI: 10.1121/1.2114607

[6] Fahy FJ. Sound and Structural Vibration: Radiation, Transmission and Response. New York: Academic; 1985

[7] Pan J, Bies DA. The effect of fluid-structural coupling on sound waves in an enclosure, theoretical part. Journal of the Acoustical Society of America. 1990;**87**(2):691-707. DOI: 10.1121/1.398939

[8] Pan J, Hansen CH, Bies DA. Active control of noise transmission through a panel into a cavity: I. Analytical study. Journal of the Acoustical Society of America. 1990;**87**:2098-2108. DOI: 10.1121/1.399555

[9] Du JT, Li WL, Liu ZG, Xu HA, Ji ZL. Acoustic analysis of a rectangular cavity with general impedance boundary conditions. Journal of the Acoustical Society of America. 2011;**130**(2):807-817. DOI: 10.1121/1.3605534

[10] Du JT, Li WL, Jin GY, Yang TJ, Liu ZG. An analytical method for the in-plane vibration analysis of rectangular plates with elastically restrained edges. Journal of Sound and Vibration. 2007;**306**:908-927. DOI: 10.1016/j.jsv.2007.06.011

[11] Li WL. Comparison of Fourier sine and cosine series expansions for beams with arbitrary boundary conditions. Journal of Sound and Vibration. 2002;**255**:185-194. DOI: 10.1006/jsvi.2001.4108

[12] Crighton DG, Dowling AP, Ffowcs Williams JE, Heckl M, Leppington FG. Modern Methods in Analytical Acoustics. Berlin: Springer-Verlag; 1992

[13] Ginsberg JH. Mechanical and Structural Vibrations: Theory and Applications. New York, John Wiley & Sons; 2001

[14] Li WL, Zhang XF, Du JT, Liu ZG. An exact series solution for the transverse vibration of rectangular plates with general elastic boundary supports. Journal of Sound and Vibration. 2009;**321**:254-269. DOI: 10.1016/j.jsv.2008.09.035

[15] Du JT, Li WL, Xu HA, Liu ZG. Vibro-acoustic analysis of a rectangular cavity bounded by a flexible panel with elastically restrained edges. Journal of the Acoustical Society of America. 2012;**131**(4):2799-2810. DOI: 10.1121/1.3693652

[16] Johnson ME, Elliott SJ, Beak K-H, Garcia-Bonito J. An equivalent source technique for calculating the sound field inside an enclosure containing scattering objects. Journal of the Acoustical Society of America. 1998;**104**:1221-1231. DOI: 10.1121/1.424330

[17] Rajalingham C, Bhat RB, Xistris GD. Natural vibration of a cavity backed rectangular plate using a receptor-rejector system. Journal of Vibration and Acoustics. 1995;**117**: 416-423. DOI: 10.1115/1.2874473

[18] Guy RW. Pressure developed within a cavity backing a finite panel when subjected to external transient excitation. Journal of the Acoustical Society of America. 1980;**68**: 1736-1747. DOI: 10.1115/1.2874473

[19] Kim SM, Brennan MJ. A compact matrix formulation using the impedance and mobility approach for the analysis of structural-acoustic systems. Journal of Vibration and Acoustics. 1999;**223**(1):97-113. DOI: 10.1006/jsvi.1998.2096

Chapter 3

Sound Waves in Complex (Dusty) Plasmas

Aamir Shahzad, Muhammad Asif Shakoori,
Maogang He and Sajid Bashir

Additional information is available at the end of the chapter

http://dx.doi.org/10.5772/intechopen.71203

Abstract

Wave properties of strongly coupled complex dusty (SCCD) plasmas evaluated using the equilibrium molecular dynamics (EMD) simulation technique. In this work, the plasma normalized longitudinal current correlation function $C_L(k,t)$ and transverse current $C_T(k,t)$ are calculated for a large range of plasma parameters of Coulomb coupling parameter (Γ) and screening strength (κ) with varying wave's number (k). In EMD simulations, we have analysed different modes of wave propagation in SCCD plasmas with increasing and decreasing sequences of different combinations of plasmas parameters (κ, Γ) at varying simulation time step (Δt). Our simulation results show that the fluctuation of waves increases with an increase of Γ and decreases with increasing κ. Additional test shows that the presented results for waves are slightly dependent on number of particles (N). The amplitude and time period of $C_L(k,t)$ and $C_T(k,t)$ also depend on different influenced parameters of κ, Γ, k and N. The new results obtained through the presented EMD method for complex dusty plasma discussed and compared with earlier simulation results based on different numerical methods. It is demonstrated that the presented model is the best tool for estimating the behaviour of waves in strongly coupled complex system (dusty plasmas) over a suitable range of parameters.

Keywords: wave properties, plasma parameter, current correlation function, dynamic structure factor, strongly coupled complex (dusty) plasmas, equilibrium molecular dynamics

1. Introduction

Transport properties of complex dusty plasma have been very actively investigated in the laboratory and by computer simulations, and wave properties have played a dominated role in both system monitoring and optimization. Waves in complex system have a dynamical role for understanding the behaviours of individual particles in different applications, for example, instabilities and wave propagation during flow. The subject of complex liquids or dusty plasmas containing micron-size charged condensed particles has recently been actively investigated in the physics and chemistry of plasmas, ionized gases, and the space environment, environmental sciences, semiconductor plasma processing industries, nuclear energy generation and materials research. Dust in atmosphere and in the entire universe is in different shapes and sizes. Mostly, it is in solid form and also in liquid and gaseous forms. Dust particle coexists with plasma and then forms dusty (complex) plasma. In the plasma when an electron is displaced from their equilibrium position, the electrostatic forces due to ions pull the electron back. Due to small mass of electron, overshoot the ions and oscillation in plasma is started.

1.1. Plasma

There are four states of matter: liquid, solid, gas and plasma. The Irving Langmuir, an American physicist, defines plasma first time as "plasma is a quasi-neutral gas of the charged particles that show collective behaviors." He received Noble prize in 1927 first time using the term "plasma" [1]. The quasi neutral means "when the number of ions is equal to the number of electrons then gas becomes electrically neutral ($ni \approx ne$)". Collective behaviour means that charged particle collides with each other due to electric field and coulomb potential. Applications of plasma in industries, science and technology and in our daily life are very significant [2, 3].

1.1.1. History of plasma

In 1922, the term plasma is used and defined by an American scientist Irving Langmuir. In 1930, a few isolated researchers, each motivated by some specific practical problems, began the study of what is now called plasma physics. In 1940, Hanes Alfven developed theory of hydrodynamic waves and gives the future prospectus of waves in astrophysical plasma (now called Alfven's waves). In 1950, research on magnetic fusion energy was started based on plasma physics in the USA, Soviet and Britain. The end decade of 1960 Russian Tokomak configuration empirically developed and created plasma with different parameters. In 1970 and 1980, the performance of tokomak was improved at the end of twentieth century and fusion break-even nearly achieved in tokomak. In the 1980s, many new applications of plasma in different fields of science and technology appeared. In the 1990s, study on dusty plasma was started (when the dust particles are immersed in plasma, they change the properties of plasma, which is known as dusty plasma) [4, 5].

1.1.2. Types of plasma

Plasma classified on the basis of temperature of electrons, ions and degree of ionization as well as density.

1.1.3. Cold plasma

T_i, T_e and T_g are the temperatures of ions, electron and gas respectively, if the plasmas satisfy $T_e >> T_i >> T_g$ condition is called cold plasma. The pressure of the cold plasma is low than collisions between electrons, ions and gases molecule that are not frequent. In this plasma, non-thermal symmetry between energies of gas molecules and electrons are not existed. Temperature of electrons is higher than ions and has greater energy than ions. Similarly, ions have greater energy than gas molecules. In this type of plasma, the effect of magnetic field is very weak and also can be ignored; only charge particles are affected by electric field. It is created in the laboratory on positive column of the glow discharge tube. Applications of cold plasma in food processing, sterilization of tooth, hand and self-decontaminating filter. It is used as a convenient descriptor to distinguish one-atmosphere. Examples of cold plasma are fluorescent lamps and neon signs [6].

1.1.4. *Ultracold plasma*

If the plasma having very low temperature approximately 100 mK to 10 μK, density $2\times10^9 cm^3$, interaction between particles is strong and the thermal energy of the particles is low compared with Coulomb energy between interacting particles known as ultracold plasma. It is created by pulsed lasers, laser-cooled atoms and photo ionization [7].

1.1.5. *Hot plasma*

If the plasmas have very high temperature, the collision between interacting particles is very frequent also thermal equilibrium obtains, if plasma satisfies $T_e \cong T_i$, this condition is known as hot plasma or thermal plasma. It is created in laboratory with high gas pressure in the discharge tube. Hot plasma acquired local thermodynamic equilibrium (LTE). Examples of hot plasma are atmospheric arcs, sparks and flame. In the hot plasma, the thermal equilibrium occurs between the gas molecules and electrons when electron collides with the gas molecules at a high pressure in the discharge tube.

1.1.6. *Ideal plasma*

It is defined by a plasma parameter which is known as Coulomb coupling and denoted by $\Gamma = \frac{P.E}{K.E}$. Whenever, $\Gamma < 1$, then that plasma is called ideal plasma. In ideal plasma, the kinetic energy (KE) of the plasma is greater than potential energy (PE) at low density and high temperature. In the ideal plasma, the value of Coulomb coupling is negligible. Due to low density, the collisions between particles are less, and ideal plasma does not have a specific structure [8, 9].

1.1.7. *Non-ideal complex (dusty) plasma*

When dust particles are immersed in the plasma, then the properties of plasma become complex known as complex (dusty) plasma. The dust particles are charged and much larger than electrons, ions and neutral atoms, and their size varies from millimetre to nanometre. The properties of dust particles are investigated in different fields of research such as plasma physics, plasma chemistry, ionized gases, material research and astrophysics and also in space physics. Dust particle changes different properties of plasma dominate in current carrier, form liquid and crystalline states. Comets, exposed dusty surfaces, planetary rings and zodiacal dust cloud are the examples of dusty plasma [10–12].

1.2. Dust particle in plasma

Dust in atmosphere and in the entire universe is in different shapes and sizes. Mostly, it is in solid form and also presents in liquid and gaseous forms. If dust particle coexists with plasma, it forms dusty (complex) plasma. Dust particle acquires charge and is affected by electric and magnetic fields. Temperature on the dust particle is approximately 10 K, and electric potential varies from 1 to 10 V is mostly negatively potential. Charge on dust particle depends on the flow of electrons and ions. It can be grown or inserted in low-temperature plasma at low degree of ionization in the laboratory. It becomes a dominant component of dusty plasma for transports of energy, momentum and mass. In these days, dusty plasma becomes more significant in the field of research of science and technology by investigating its transport properties in the laboratory. Currently, complex plasma becomes more interesting field of research. The major ratio of plasma in the universe is dusty plasma. It is a huge occurrence in magneto and ionosphere of earth, atmosphere of stars, sun, solar wind, galaxies cosmic radiations, planetary rings comet tails and nebulae. Dusty plasma also known as complex plasma has electrons, ions, neutral and dust particles components [12].

1.2.1. Charge on dust particle

Dust particles in plasma gain the electric charge, which they obtained in the plasma. The charge on dust particle can range from zero to hundreds of thousands of electrons depending on the size and shape of the dust particle. When a dust grain immersed in an ionized medium, soon it gains a charge when its surface is contact with plasma. This is valid for small floating objects like dust particles, also for electrode surfaces and macroscopic objects inserted into the discharge tube. This charge determines by balancing positive ion and the electron current. These currents must be equal in a steady state. Due to high mobility of the electrons as compared to ions, the surface collects negative charges, which attracts positive ions and repel the electrons until an immobile state is reached. Therefore, by equating fluxes between electrons and ions, the charge on particle can be obtained. This charge is responsible for confinement and long lifetimes of particles in plasmas. Due to interactions between these particles, waves, instabilities and other collective phenomena are produced.

1.2.2. Size of dust particle

Dust particles are much larger than electrons and ions, and its size varies from 10^{-3} m to 10^{-9} m. A typically used particle may have a diameter of 3.50±0.05 μm and a mass ~3×10^{-11} kg. Such particles are named mono disperse. It has any shape, mostly spherical, and also easily observed in laboratory without any microscope. It is made by conducting or dielectric material [12, 13].

1.2.3. Nature of dust particle

Dust particles that collect the plasma particles such as electrons and ions on their surface become an interacting particle. In the dusty plasma energy, momentum and mass are transferred through dust particle. Dust particle may be in liquid, solid and gaseous forms.

1.3. Application of plasma in daily life

Plasma is widely used in the field of science, technology and in our daily life. Plasma science also affects human life in different fields and ways. It plays a very important role in the sterilizing of medical instruments, laser, developments of fusion energy, intense particle beams, plasma processing, lightening, development fusion energy controlling, high-power energy sources, high-power radiations sources and water purification plant [3]. X-ray and ultraviolet electron beams and radiations, which are emitted by plasma-centered sources, have a diversity of applications in different fields.

1.3.1. Application of plasma in textiles

In textile industries, plasma used for surface treatment shows a large number of advantages. Different plasma treatments were used to modify the surface of fabrics, such as cold plasma treatments, which modify the large and reliable systems in fabrics. Three effects are found on the textile surfaces during plasma treatments: (1) cleaning effect, (2) increase of micro roughness and (3) produce deliquescent surfaces. The plasma polymerization is a process in which solid polymeric materials are deposited on textile substrate with required properties; currently, this technique is under development, and in this technique only upper surface is modified. This treatment also has optimistic effects on printing and dying of wool and has proved to be successful in shrink-resist. Improvement of surface wetting technique in artificial polymers is done with treatment in oxygen-, air- and NH_3 plasma. Hydrophilic treatment used as a dirt-repellent and antistatic finish [Sparavigna, 2008]. Plasma treatment is environmental friendly in which only upper layer of surface is modified without changing its bulk properties. Vacuum plasma is used on a small scale as is used in reel-to-reel machinery at the industrial scale; it plays a vital role to enhance a number of properties, including liquid repellency, the specific surface functionality and improved wet ability.

1.3.2. Application of plasma in industry

The surfaces of optics and contact lenses can be cleaned by plasma treatment. In this process, thin layer of the organic impurity is removed almost from all surfaces. Plasmas are used for the manufacturing of semiconductor chips in discharge tube, and the dust particles are essentially formed from the reactive gases used for the creation of plasmas. These dust particles are used for the manufacturing of semiconductor chips and give the significant results for the conduction of electric current. In the aerospace industries, plasma treatment of composite and polymer materials are commonly used. This treatment has advantages in adhesive and cleaning bonding of the dissimilar materials. In the automotive industries, adhesion of the exterior and interior parts is increased by the plasma treatment apparatus and plasma treatment procedures without any heat effect that commonly facing with flames. The Henniker is well known as a great leader in this field because he controlled plasma systems with low cost for almost in every common automotive development mission [2, 3, 11].

1.4. Waves in plasma

1.4.1. Plasma oscillation

In the plasma, when an electron is displaced from their equilibrium position, the electrostatic forces due to ions pull the electron back. Due to small mass of electron, overshoot the ions and oscillation in plasma are started. The plasma oscillation leads to plasma frequency which is denoted by ω_{pe}. When the amplitude of oscillation is small, then oscillation forms a sinusoidal waveform. Mathematically, oscillation is represented as:

$$n = \overline{n}\exp[i(k.r - \omega t)] \tag{1}$$

where n is the density, $\overline{n}$ is the constant that tells about the amplitude of waves and k is propagation constant that describes the direction of waves and t is the time. The propagation constant in Cartesian coordinates

$$k.r = k_x x + k_y y + k_y y \tag{2}$$

The phase velocity is $\frac{dx}{dt} = \frac{\omega}{k} = v_\phi$ that gives the direction of waves negatively or positively. Sometimes the phase velocity in plasma exceeds the velocity of light c. The group velocity is $v_g = \frac{dw}{dk}$ and plasma frequency is $\omega_{pe} = \left(\frac{n_o e^2}{\varepsilon_o m}\right)^{\frac{1}{2}}$ that depends upon the density of plasma and does not depend on k.

1.4.2. Classification of waves in plasma

In this chapter, we classified plasma waves with respect to direction of motion of charge particles (electron, ion), which are perpendicular or parallel to the direction of propagation. The longitudinal waves consist of electrostatic (ion and electron) waves and transverse wave consists of electromagnetic (ion and electron) waves, mostly transverse component and sometime longitudinal component also.

1.5. Electron waves (electrostatic)

1.5.1. Plasma oscillation

Plasma oscillation, also known as Irving Langmuir waves, is discovered by American scientists Irving Langmuir and Lewi Tonks in 1920, and the frequency of plasma oscillation is high due to low electron density in the conducting materials such as metal and plasma. It is found in interstellar gas clouds and earth atmosphere. Mathematically, it is written as $\omega^2 = \omega_p^2 + \frac{3}{2} k^2 v_{th}^2$.

1.5.2. Upper hybrid oscillation

The upper hybrid oscillation is a resonance phenomenon in plasma which the electric field perpendicular to the magnetic field and waves propagates across the static magnetic field: $\omega^2 = \omega_p^2 + \omega_c^2 = \omega_h^2$.

1.6. Ions waves (electrostatic)

1.6.1. Acoustic waves

It is a type of longitudinal waves in magnetized plasma that propagates due to compression mode and its direction along the magnetic field lines. The particle velocity, sound pressure, sound intensity and particle displacement are the important quantities of acoustic waves. It exhibits the interference, diffraction and reflection phenomenon. In general, ion acoustic wave formula is [1, 14], $\omega^2 = k^2 v_s^{\ 2} = k^2 \frac{\gamma_e KT_e + \gamma_i KT_i}{M}$.

2. EMD model for correlation relations and numerical technique

In this section, we introduce the molecular dynamic simulation technique and its theoretical background that was needed in this work. We introduced a system of mathematical model that used and applied in the computer simulations and also explained steps for molecular dynamics simulation codes. In addition, it also explains the simulation parameters and techniques. Yukawa potential was first time proposed by a scientist Hideki Yukawa in 1930. Through this potential, charged particles do interact, which is defined as

$$\phi(|r|) = \frac{Q^2}{4\pi\varepsilon_o} \frac{e^{-|r|/\lambda_D}}{|r|} \tag{3}$$

where Q is the charge on dust particle interacts with other charge particle, ε_0 is the permittivity of free space, r is the distance of interacting particle and λ_D is Debye screening length.

2.1. Current correlation functions

The particle current or momentum current for single atomic (molecule) species in MD unit is given as:

$$\pi(r, t) = \sum_j v_j \delta\left(r - r_j(t)\right) \tag{4}$$

where v_j and r_j are the velocity and position of jth particle. By using the Fourier transformation, Eq. (4) should be written as:

$$\pi(k, t) = \sum_j v_j e^{-ik.r_j(t)} \tag{5}$$

The correlation function of the current vector component is defined as:

$$C_{\alpha\beta}(k, t) = \frac{k^2}{N_m}\Big(\pi_\alpha(k, t)\, \pi_\beta(-k, 0) \tag{6}$$

For the isotropic fluid under the consideration of symmetry, Eq. (6) can be expressed in terms of longitudinal current correlation and transverse current correlation in the relative direction of k, where k is the wave vector.

$$C_{\alpha\beta}(k, t) = \frac{k_\alpha k_\beta}{k^2} C_L(k, t) + \left(\delta_{\alpha\beta} - \frac{k_{\alpha\beta}}{k^2}\right) C_T(k, t) \tag{7}$$

By setting $k = k\hat{z}$, we obtain longitudinal and transverse current functions in X, Y and Z direction relative to the k and $-k$.

2.2. Normalized longitudinal current correlation function $C_L(k,t)$

Eq. (7) split into longitudinal current and transverse current by using the Fourier transformation as:

$$C_L(k, t) = \frac{k^2}{N_m} \langle \pi_z(k, t)\, \pi_z(-k, 0) \rangle \tag{8}$$

Inserting the value of π_z in Eq. (8), we have

$$C_L(k, t) = \frac{k^2}{N_m} \langle \Sigma\, v_i\, e^{-ik.z_i(t)}(k, t) \Sigma v_j\, e^{-ik.Z}(k, t) \rangle \tag{9}$$

Longitudinal current correlation function explains the direction of propagates along Z-axis in the positive and negative directions of Z-axis with wave number. This equation also specifies about longitudinal motion of charge particle.

2.3. Time-dependent normalized transverse current correlation $C_T(k,t)$

Using the Fourier transformation for Eq. (7), the transverse current correlation is defined as:

$$C_T(k, t) = \frac{k^2}{2\,N_m} \langle \pi_x(k, t)\, \pi_x(-k, 0) + \pi_y(k, t)\, \pi_y(-k, t) \rangle \tag{10}$$

where

$$\pi_x(k, t) = \sum_j v_j\, e^{-ik.x_j(t)} \tag{11}$$

$$\pi_y(k, t) = \sum_j v_j\, e^{-ik.y_j(t)} \tag{12}$$

Eq. (12) states about the directions of propagation of electromagnetic waves with respect to wave number along the positive and negative X- and Y-axes. The transverse current correlations also describe us about transverse motion of charge particle in complex system or complex (dusty) plasma [15–17].

2.4. Parameters and simulation techniques

In this section, we select a system of choice having number of particles $N = 500$–4000: these existing particles in a cube volume (V) interact with each other by pairwise Yukawa potential given in Eq. (3). In this chapter, the EMD simulation technique has been used to investigate the longitudinal current correlation and transverse current correlation that are given in Eqs. (9) and (10), respectively. The dimensions of simulation box are chosen as $L_{x,}$ $L_{Y,}$ L_Z. The periodic boundary condition is used to minimize the surface size effect and applied to the simulation box. In our case, the main calculations are performed for $N = 500$ particles at $\kappa = 1.4$, 2, 3 and 4 with plasma coupling parameters Γ (temperature of the Yukawa system) varies from 1 to 300 and waves numbers $k = 0$, 1, 2 and 3. The simulation time step is taken as $\Delta t = 0.005/\omega_p$ that allows to compute the important data for sufficient long simulation run. The EMD method is reported for the current correlation of SCCD plasma over reasonable domain of plasma parameters of Debye screening ($1 \leq \kappa \leq 4$) and Coulomb coupling ($1 \leq \Gamma \leq 300$). Particle velocity thermostat is used to control the temperature of systems. The reported simulations are performed between $5.0 \times 10^5/\omega_p$ and $2.0 \times 10^5/\omega_p$ time units in the series of data recording of correlation functions [16, 17].

3. EMD simulation results

3.1. Time-dependent longitudinal current correlation function $C_L(k,t)$, at $\kappa = 1.4$

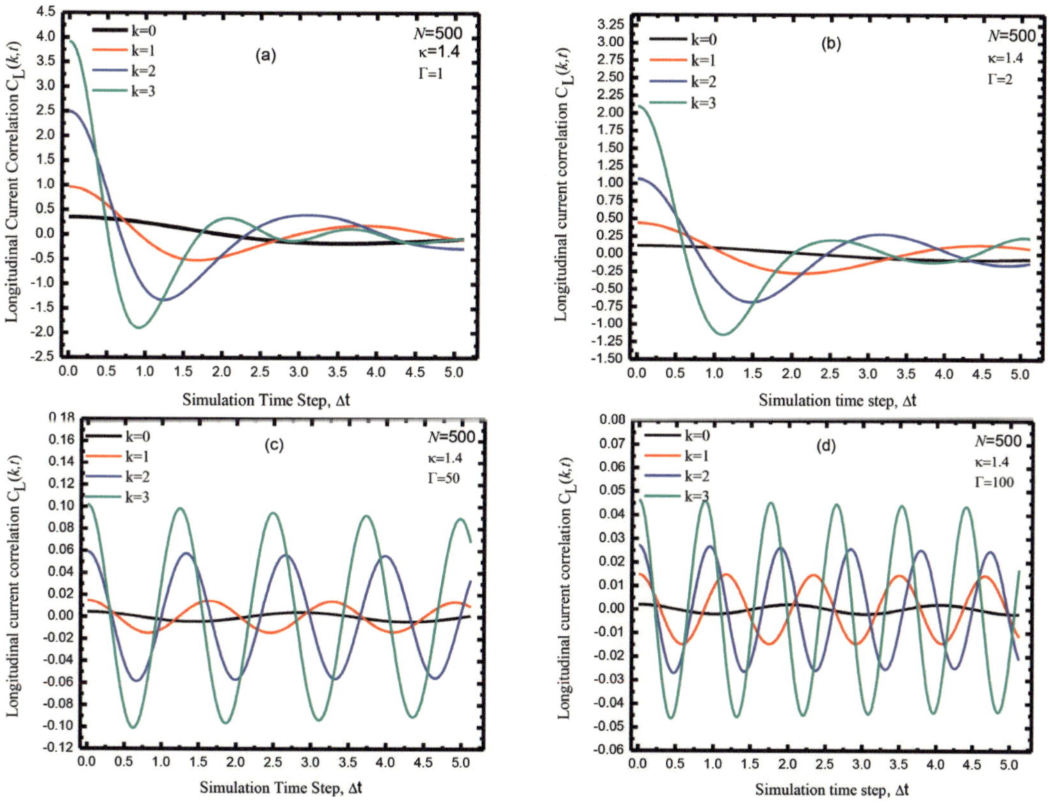

Figure 1. Variation of $C_L(k,t)$ as a function of simulation time step (Δt) of strongly coupled complex plasma at $\kappa = 1.4$, $N = 500$, waves number $k = 0$, 1, 2 and 3 for (a) $\Gamma = 1$, (b) $\Gamma = 2$, (c) $\Gamma = 50$ and (d) $\Gamma = 100$.

3.2. Time-dependent longitudinal current correlation function $C_L(k,t)$ at $\kappa = 2$

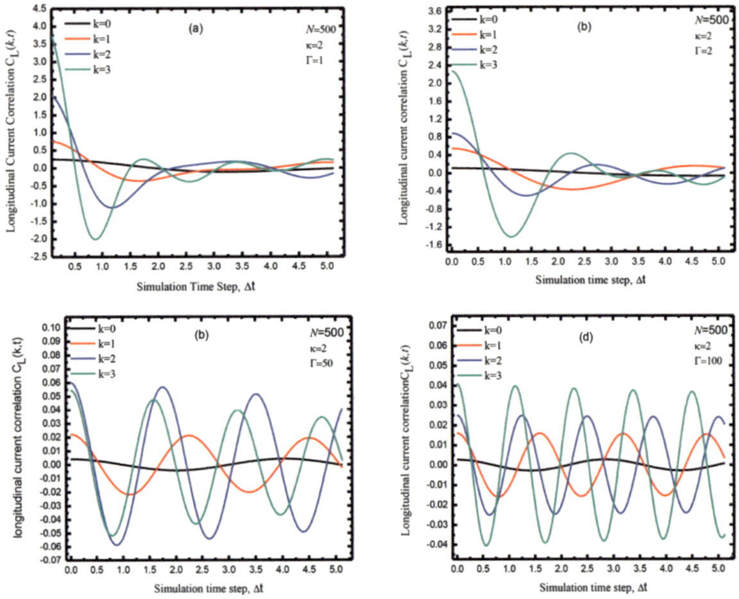

Figure 2. Variation of $C_L(k,t)$ as a function of simulation time step (Δt) of strongly coupled complex plasma at κ =2, N = 500, waves number k = 0, 1, 2 and 3 for (a) Γ =1, (b) Γ =2, (c) Γ =50 and (d) Γ =100.

3.3. Time-dependent longitudinal current correlation function $C_L(k,t)$ at $\kappa = 3$

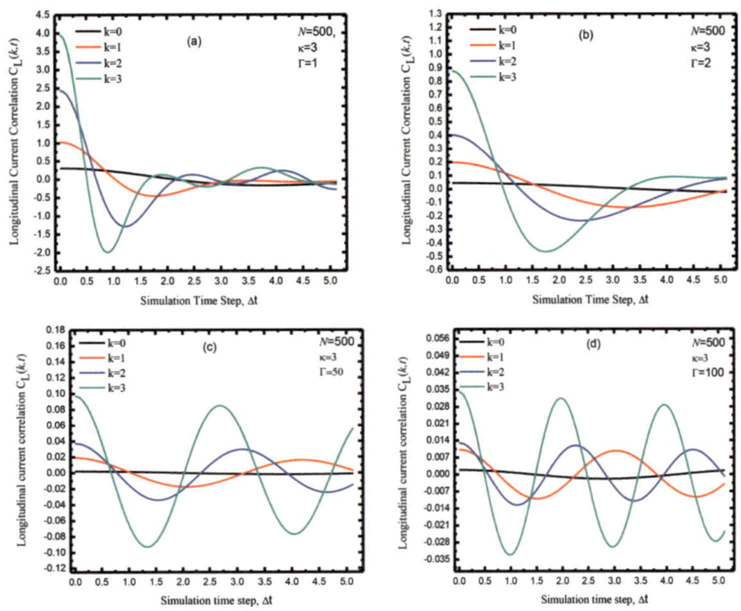

Figure 3. Variation of $C_L(k,t)$ as a function of simulation time step (Δt) of strongly coupled complex plasma at κ =3, N = 500, waves number k = 0, 1, 2 and 3 for (a) Γ =1, (b) Γ =2, (c) Γ =50 and (d) Γ =100.

http://dx.doi.org/10.5772/intechopen.71203

To illustrate the behaviour of correlation function of SCCD plasma using EMDS, the time-dependent normalized longitudinal current correlation ($C_L(k,t)$ from Eq. (9) spectra is plotted against simulation time step (Δt), as shown in **Figures 1–3**, for waves number (k = 0, 1, 2 and 3), plasma parameter (Coulomb coupling) Γ = 1, 2, 50 and 100 and for screening parameter (κ = 1.4, 2 and 3), respectively. It is observed from four panels of **Figure 1** that the wavelength decreases and amplitude of longitudinal current increases with increasing waves numbers (k), as $C_{L=}$ 0.2875, 0.9220, 2.5143 and 3.7139 for waves numbers k = 0, 1, 2 and 3, respectively, at same values of Γ =1 and κ =1.4. It has been shown that the value of $C_L(k,t)$ decreases with increasing of Γ as $C_L(k,t)$ = 0.2875 (0.9220), 0.1179 (0.4369), 0.0044 (0.0148) and 0.0020 (0.0148) for Γ =1, 2, 50 and 100, respectively, at k = 0 (1) and for κ =1.4. In case of wave number k = 2(3), the value of C_L is 2.5142(3.7139), 1.0661(2.0954), 0.0591(0.1020) and 0.0271(0.0463), respectively. The fluctuation of sound waves in SCCD plasma is observed from **Figures 1–3** that increases with decreasing the plasma temperature, and it is noted that the amplitude, fluctuation and wavelength for simulation time step (Δt) of longitudinal current correlation in SCCD plasma gradually decrease with increasing of κ. We have noted from simulation results that the propagation of sound waves in dusty plasma is frequently propagated at higher value of Γ =50 and 100 and at lower value of κ =1.4 and 2.

3.4. Time-dependent transverse current correlation function $C_T(k,t)$, κ = 1.4

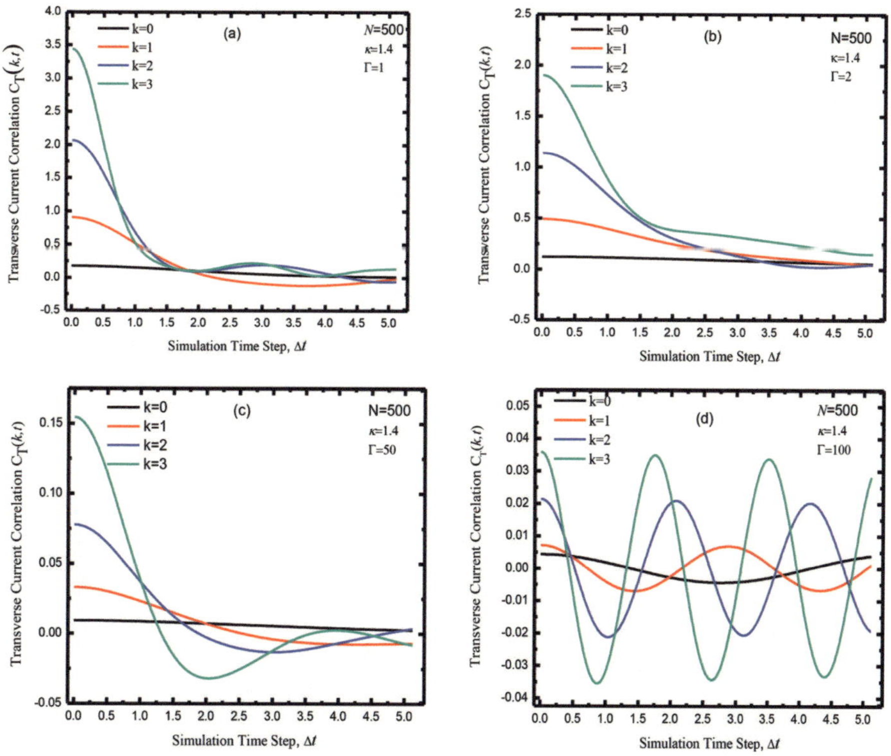

Figure 4. Variation of $C_T(k,t)$ as a function of simulation time step (Δt) of strongly coupled complex plasma at κ =1, N = 500, waves number k = 0, 1, 2 and 3 for (a) Γ =1, (b) Γ =2, (c) Γ =50 and (d) Γ =100.

3.5. Time-dependent transverse current correlation function $C_T(k,t)$, $\kappa = 2$

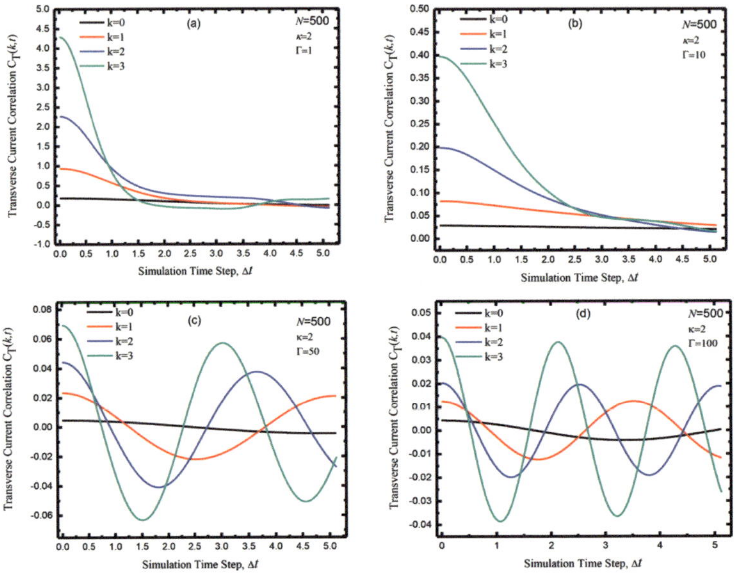

Figure 5. Variation of $C_T(k,t)$ as a function of simulation time step (Δt) of strongly coupled complex plasma at κ =2, N = 500, waves number k = 0, 1, 2 and 3 for (a) Γ =1, (b) Γ =2, (c) Γ =50 and (d) Γ =100.

3.6. Time-dependent transverse current correlation function $C_T(k,t)$, $\kappa = 4$

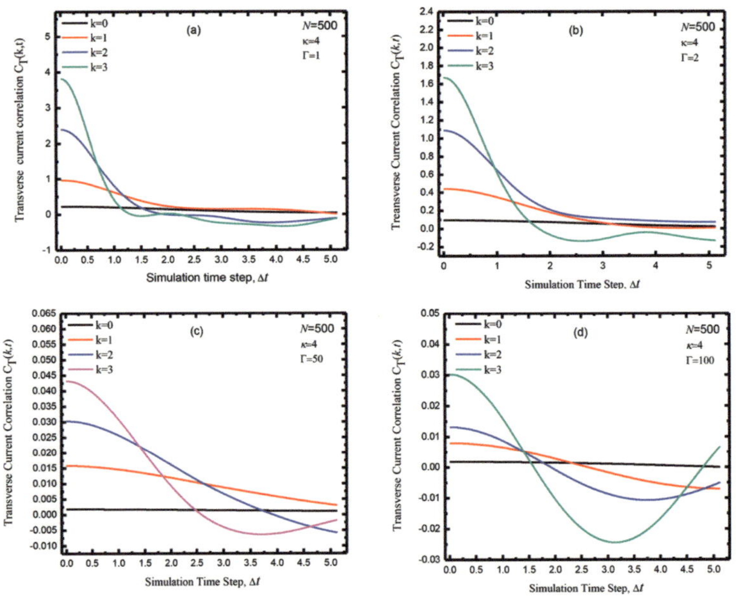

Figure 6. Variation of $C_T(k,t)$ as a function of simulation time step (Δt) of strongly coupled complex plasma at κ =4, N = 500, waves number k = 0, 1, 2 and 3 for (a) Γ =1, (b) Γ =2, (c) Γ =50 and (d) Γ =100.

Figures 4–6 demonstrate that the simulation results are obtained for $C_T(k,t)$ of SCCD plasma using EMD simulation at $k = 0, 1, 2$ and 3, $\Gamma =1, 2, 50$ and 100 and $\kappa =1.4, 2$ and 4, respectively, at $N = 500$. The presented simulation results of $C_T(k,t)$ spectra are compared with increasing and decreasing sequence of Γ, κ and k. It is noted that the fluctuation, wavelength and amplitude of $C_T(k,t)$ (transverse or shear waves) in complex (dusty) plasma immediately depend on plasma temperature, Coulomb coupling and system size. We have observed from simulation results of transverse waves in dusty plasma that wavelength of these waves decreases with increasing wave number, plasma parameters, Coulomb coupling and number of particles. The amplitude of $C_T(k,t)$ increases with increasing wave number as $C_T = 0.2655, 0.9679, 2.0914$ and 3.8960 for $k = 0, 1, 2$, and 3, respectively, at $\kappa =1.4$ shown in **Figure 4(a)** and decreases with increasing Γ as for $k = 0(1)$, $C_T = 0.2655(0.9679), 0.1395(0.5607), 0.0074(0.0197)$ and $0.0040(0.0081)$ and for the case $k = 2(3)$, $C_T = 2.0914(3.8960), 1.0590(1.8826), 0.0537(0.0929)$ and $0.0259(0.0366)$ at $\Gamma =1, 2, 50$ and 100, respectively, shown in four panels of **Figures 4** at $\kappa =1.4$ and $N = 500$. We have observed from 12 simulated results, which are displayed in **Figures 4–6**, that fluctuations in $C_T(k,t)$ increase with decreasing plasma temperature and decrease with increasing κ, N. We have concluded from 24 simulated results that amplitude and wavelength of transverse waves in SCCD plasma are directly proportional to the plasma temperature and inversely proportional to the screening strength. Shear waves are frequently propagated at higher value of Γ.

4. Summary

The EMD simulation method is used to investigate the $C_L(k,t)$ and $C_T(k,t)$ for SCCD plasma over a wide range of plasma parameters κ, Γ, N and k (wave number). The first involvement of presented simulation is that it delivers and understands the propagation of waves in SCCD plasma. In general, the amplitude and frequency of waves with different modes are analysed. The presented simulation specifies that the waves are frequently propagated at intermediate and high value of Γ. These investigations show that the values of frequency and amplitude depend on Γ, κ, N and k. The EMD simulation investigation shows the plasma wave behaviors that are observed at low and intermediate values of Γ and does not show at higher value of Γ and at lower value of $\kappa =1.4$. It has been shown that the presented EMD method and earlier EMD techniques have comparable performance over the wide range of plasma points, both yielding reasonable results for correlation parameters. New simulations yield more reliable and excellent data for the $C_L(k,t)$ and $C_T(k,t)$ for a wider range of κ (= 1, 4) and Γ (= 1, 100), respectively. The existing simulation delivers more reliable data for the existence of waves and dynamical structure in SCCD plasma. These simulation results show that in the absence of structure in dusty plasma, the shear wave does not support. The sound wave frequently propagates at medium and higher values of Γ in SCCD plasma. It is suggested that the presented EMD technique based on the Ewald summation described here can be used to explore other ionic and dipolar materials. It is very interesting to note what other types of interaction potentials support correction parameters of strongly coupled plasmas and how its strengths depend on the range of new interaction potentials.

Acknowledgements

The authors thank Z. Donkó (Hungarian Academy of Sciences) for providing his thermal conductivity data of Yukawa Liquids for the comparisons of our simulation results and useful discussions. We are grateful to the National High Performance Computing Center of Xian Jiaotong University and National Advanced Computing Center of National Center of Physics (NCP), Pakistan, for allocating computer time to test and run our MD code.

Abbreviations

NICDPs	non-ideal complex (dusty) plasmas
EMD	equilibrium molecular dynamics
Γ	Coulomb coupling
κ	Debye screening length
$C_T(k,t)$	transverse current correlation function
MD	molecular dynamics
LTE	local thermodynamic equilibrium
SCP	strongly coupled plasma
PBCs	periodic boundary conditions
$C_L(k,t)$	longitudinal current correlation function

Author details

Aamir Shahzad[1,2]*, Muhammad Asif Shakoori[1], Maogang He[2] and Sajid Bashir[3]

Address all correspondence to: aamirshahzad_8@hotmail.com

1 Department of Physics, Government College University Faisalabad (GCUF), Faisalabad, Pakistan

2 Key Laboratory of Thermo-Fluid Science and Engineering, Ministry of Education (MOE), Xi'an Jiaotong University, Xi'an, PR China

3 Punjab Institute of Nuclear Medicine Faisalabad, Faisalabad, Pakistan

References

[1] Chen FF. Introduction to Plasma Physics and Controlled Fusion. 2nd ed. New York: Springer-Verlag. 2006; 200 p. ISBN: 9780521825689

[2] Shahzad A, He M-G. Thermal conductivity calculation of complex (dusty) plasmas. Physics of Plasmas. 2012;**19**(8):083707. DOI: 10.1063/1.4748526

[3] Shahzad A, He M-G. Computer Simulation of Complex Plasmas: Molecular Modeling and Elementary Processes in Complex Plasmas. 1st ed. Saarbrücken, Germany: Scholar's Press; 2014; 170 p

[4] Bellan PM. Fundamentals of Plasma Physics. 1st ed. Cambridge, UK: Cambridge University Press; 2008; ISBN: 9780521528009

[5] Shahzad A, He M-G. Thermal conductivity of three-dimensional Yukawa liquids (dusty plasmas). Contributions to Plasma Physics. 2012;**52**(8):667. DOI: 10.1002/ctpp.201200002

[6] Killian T, Pattard T, Pohl T, Rost J. Ultracold neutral plasmas. Physics Reports. 2007;**449**:77. DOI: 10.1016/j.physrep.2007.04.007

[7] Killian TC, Kulin S, Bergeson SD, Orozco LA, Orzel C, Rolston SL. Creation of an ultracold neutral plasma. Physical Review Letters. 1999;**83**(23):4776. DOI: 10.1103/phyRevLett. 83.4776

[8] Boulos MI, Fauchais P, Pfender E. Thermal Plasmas: Fundamentals and Applications. Vol. 8, New York: Springer Science & Business Media; 2013. DOI: 10.1007/978-1-1337-1

[9] Shahzad A, He M-G. Thermal Conductivity and Non-Newtonian Behavior of Complex Plasma Liquids, chapter from book of Thermoelectrics for Power Generation—A Look at Trends in the Technology" InTech. 2016; 305 p. DOI: 10.5772/65563

[10] Fortov VE, Vaulina OS, Lisin EA, Gavrikov AV, Petrov OF. Analysis of pair interparticle interaction in nonideal dissipative systems. Journal of Experimental and Theoretical Physics. 2010;**110**:662-674. DOI: 10.1134/S1063776110040138

[11] Shahzad A, He M-G. Thermodynamic characteristics of dusty plasma studied by using molecular dynamics simulation. Plasma Science and Technology. 2012;**14**(9):771-777. DOI: 10.1088/1009-0630/14/9/01

[12] Shahzad A, He M-G. Interaction contributions in thermal conductivity of three dimensional complex liquids. In: Liejin GUO, editor. AIP Conf. Proc; 26-30-10-2012; Xi'an, China. USA: AIP. 2013; p. 173-180. DOI: 10.1063/1.4816866

[13] Shukla PK, Mamun AA. Introduction to Dusty Plasma Physics. CRC Press; 2015 97807 50306539 – CAT#IP391

[14] Donko Z, Kalman GJ, Hartmann P. Dynamical correlations and collective excitations of Yukawa liquids. Journal of Physics: Condensed Matter. 2008;**20**(41):413101. DOI: 10.1088/0953-8984/20/41/413101

[15] Rapaport DC. The Art of Molecular Dynamics Simulation. New York: Cambridge University Press; 2004; 250 p. DOI: 9780521825689

[16] Shahzad A, He M-G. Homogenous nonequilibrium molecular dynamics evaluations of thermal conductivity 2D Yukawa liquids. International Journal of Thermophysics. 2015;**36**(10-11):2565. DOI: 10.1007/s10765-014-1671-8

[17] Shahzad A, He M-G. Calculations of thermal conductivity of complex (dusty) plasmas using homogenous nonequilibrium molecular simulations. Radiation Effects and Defects in Solids. 2015;**170**(9):758-770. DOI: 10.1080/10420150.2015.1108316

Chapter 4

Acoustic Wave Monitoring of Fluid Dynamics in the Rock Massif with Anomaly Density, Stressed and Plastic Hierarchic Inclusions

Olga Hachay and Andrey Khachay

Additional information is available at the end of the chapter

http://dx.doi.org/10.5772/intechopen.70590

Abstract

The geological environment is an open system, on which external and internal factors act. They lead it to an unstable state, which, as a rule, manifests itself locally in the form of zones, called dynamically active elements, which are indicators of potential catastrophic sources. These objects differ from the host geological environment by structural forms, which are often forming of a hierarchical type. The process of their activation can be observed using monitoring with wave fields, for mathematical support of which new modeling algorithms have been developed using the method of integral and integral-differential equations. A new approach to the interpretation of wave fields has been developed, to determine contours or surfaces of locally stressed hierarchical objects. An iterative process of solving the theoretical inverse problem for the case of determining configurations of 2D hierarchical inclusions of the k-th rank is developed. When interpreting monitoring results, it is necessary to use data from such monitoring systems that are tuned to study the hierarchical structure of the environment.

Keywords: hierarchical structures, acoustic wave fields monitoring, anomaly density, stressed, plastic, fluid saturated hierarchic inclusions, equation of theoretical inverse problem

1. Introduction

It is known that the geological environment is an open dynamic system that is influenced by natural and artificial effects at various scale levels that change its state, resulting to a complex multi-ranked hierarchical evolution. That is one of the subjects of geosynergetic study [1], see the bibliography. Using the synergetic approach, it is necessary to clear distinguish the scale of natural phenomena. The paradigm of physical mesomechanics introduced by academician

Panin V. and his school, which includes the synergetic approach, is a constructive tool for studying and changing the state of heterogeneous materials.

Determining the state and its dynamics of the massif is a more complex problem than mapping its structure. Individual parts of the massif can be in a different stress state, and the corresponding deformations can be either elastic or plastic. The medium can be multiphase. A sharp change of the state of the blocks can lead to loss of stability of the whole massif and to a rock shock. The change of the state is determined both by natural and technogenic influence on the massif and manifests itself by the formation of man-made cavities and pumping by mechanical energy during mass explosions provided for mining technologies. The phenomenon of non-stationary state of the rock massif for today is a well-known fact [2, 3]. Manifestations of it can be from insignificant ones in the form of increase in permeability due to increase in fracturing of rocks, which are already registered in the form of shocks, micro-impacts, rock impacts, mountain-tectonic impacts [4]. The latter refer already to catastrophic phenomena, which are initiated by both internal and external technogenic causes. In our studies of the non-stationary geological environment in the framework of natural experiments in real rock massive under strong man-made influence, it was shown that the dynamics of the state can be detected using synergetic effect in hierarchical environments. An important role for the study of the state of dynamic geological systems consists on a combination of active and passive geophysical monitoring, which can be carried out using electromagnetic and seismic fields. The change of the system state on the investigated spatial bases and times is manifested in parameters related to the structural features of the medium of the second and higher ranks. Thus, the study of the state dynamics, its structure and the phenomenon of self-organization of the massif should be studied by geophysical methods tuned to the multi-rank hierarchical nonstationary model of the environment.

In the work [1] the results of the method of studying synergetic manifestations of a geological environment under active external influence were generalized using the method of phase portraits for researching the problem of reflecting the synergetic properties of a geological medium in the data of active electromagnetic and seismic monitoring. The results, obtained from the analysis of the detailed seismological catalog from the point of view of the mathematical foundations of synergetics and open dynamical systems possessing the properties of nonlinearity and dissipativity lead us to the necessity of formulating a new mathematical modeling problem different from the one previously performed [5, 6].

The processes of development of oil and gas fields are associated with the motion of multiphase multicomponent media that are characterized by no equilibrium and non-linear rheological properties. The real behavior of reservoir systems is determined by the complexity of the rheology of moving fluids and the morphological structure of the porous medium, as well as by the variety of processes of interaction between the liquid and the porous medium [7]. Accounting for these factors is necessary for a meaningful description of filtration processes due to nonlinearity, non-equilibrium and in homogeneity inherent in real systems. In this case, new synergetic effects are revealed (loss of stability with the appearance of oscillations, the formation of ordered structures). This allows us to propose new methods for monitoring and managing complex natural systems that are tuned to account for these phenomena. Thus, the reservoir system from which to extract oil is a complex dynamic hierarchical system.

A major result of studies of the last century was the conclusion about the fundamental role of the block-hierarchical structure of rocks and massive for explaining the existence of a wide range of nonlinear geomechanical effects and the emergence of complex self-organizing geosystems in the analysis of the formation of large and super-large deposits. The hierarchical structure is characteristic for many systems, especially for the Earth's lithosphere, where more than 30 hierarchical levels from tectonic plates with a length of thousands of kilometers to individual mineral grains of millimeter size were identified by geophysical studies. Thus, the Earth's crust is a continuous medium that includes a discrete block system and, like any synergetic discrete ensemble, has hierarchical and self-similar properties. This must be taken into account when creating new complex geophysical systems for studying the Earth's lithosphere. Iterative algorithms of 2-D modeling for sound diffraction and a linearly polarized transverse elastic wave on inclusion with a hierarchical elastic structure located in the J-th layer of the N-layer elastic medium are constructed.

In the present chapter we consider the case where the inclusion density of each rank differs from the density of the enclosing medium, and the elastic parameters coincide with the elastic parameters of the enclosing layer. We consider also the case when the inclusion density of each rank coincides with the density of the host medium, and the elastic parameters of the inclusion of each rank differ from the elastic parameters of the enclosing layer. We used the method of integral and integral-differential equations for the spatial-frequency representation of the distribution of wave fields. It follows from the constructed theory that when combining acoustic, geomechanical and gravitational fields, it is necessary to use such data that are obtained within the framework of observation systems that are tuned to the study of the hierarchical structure of the medium. The use of density values obtained from the correlation between the values of the longitudinal wave velocity determined from the kinematic interpretation of seismic data and the density to construct a density model for gravity data may lead to a mismatch between this model and the real composition of the geological medium studied. The use of values of elastic parameters without taking into account density anomalies can lead to a discrepancy between the geomechanical model and the acting stresses in the geological environment.

The algorithm developed in [8] is based on Hooke's law [9]. Equations of motion are obtained by equating the elastic forces of the products of masses with accelerations, and the action of the other forces is not assumed. This assumption is completely justified for small deformations and quite often agrees well with the experimental data. However, if vibrations occur in the medium, then part of the elastic energy passes into heat due to internal friction. At the present time, the theory of internal friction in solids is developing [9]. There are several indirect methods of determining internal friction that arise in the samples, which are associated with the assumption that the restoring forces are proportional to the amplitude of the oscillation, and the dissipative forces are proportional to the velocity.

The present work is devoted to the developing an algorithm for the propagation of the seismic field in the acoustic approximation in a layer-block elastic medium with a hierarchical plastic inclusion (the case of taking into account internal friction in viscoelastic inclusion), with anomaly density, anomaly stressed and fluid-saturated hierarchical inclusion.

2. Algorithms of mathematical modeling of acoustic wave distribution in block layered hierarchical structure with different physical properties

2.1. Modeling of sound diffraction on 2D plastic hierarchic heterogeneity, located in *N*-layered elastic medium

In the book [9], considering the motion equation:

$$P = M\ddot{\xi} + \eta\dot{\xi} + E\xi \tag{1}$$

for oscillating body it was made an assumption, that the elastic reducing force E is proportional to the displacement ξ, and the dissipative force is proportional to the velocity $\dot{\xi}$, by that E in (1) depends from the elastic constants, and η depends from the dissipative forces, the nature of these in [9] did not been discussed. Let us consider the Voigt model [9] for an elastic-viscous environment that in contrary of the Hookers model introduces in the relations for the elastic constants of Lame such expressions:

$$\lambda + \lambda^{'}\omega_1;\ \mu + \mu^{'}\omega_2 \tag{2}$$

In the paper [10] it is described the algorithm of sound diffraction modeling on elastic 2D homogeneous inclusion, located in the *J*-th layer of the *N*-layered medium.

$G_{Sp,j}(M,M^0)$ – a source function of seismic field, the boundary problem of it is formulated in the paper [10], $k_{1ji}^2 = \omega^2(\sigma_{ji}/\lambda_{ji})$ – wave number for the longitudinal wave, in the cited expressions index ji identifies the medium properties inside the inclusion, ja – outside the inclusion, λ – constant of Lame, σ – medium density, ω – cycle frequency, $\vec{u} = grad\varphi$ – displacement vector, φ^0 – potential of the normal seismic field inside the layered medium without the inclusion: $\varphi_{ji}^0 = \varphi_{ja}^0$.

$$\begin{aligned}
&\frac{\left(k_{1ji}^2 - k_{1j}^2\right)}{2\pi}\iint\limits_{Sc}\varphi(M)G_{Sp,j}\left(M,M^0\right)d\tau_M + \frac{\sigma_{ja}}{\sigma_{ji}}\varphi^0\left(M^0\right) - \\
&-\frac{(\sigma_{ja} - \sigma_{ji})}{\sigma_{ji}2\pi}\oint_c G_{Sp,j}\frac{\partial\varphi}{\partial n}dc = \varphi\left(M^0\right), M^0 \in S_c \\
&\frac{\sigma_{ji}\left(k_{1ji}^2 - k_{1j}^2\right)}{\sigma\left(M^0\right)2\pi}\iint\limits_{Sc}\varphi(M)G_{Sp,j}\left(M,M^0\right)d\tau_M + \varphi^0\left(M^0\right) - \\
&-\frac{(\sigma_{ja} - \sigma_{ji})}{\sigma\left(M^0\right)2\pi}\oint_c G_{Sp,j}\frac{\partial\varphi}{\partial n}dc = \varphi\left(M^0\right), M^0 \notin S_c.
\end{aligned} \tag{3}$$

Let us use the expression (2), $k_{1ji}^2 = \omega^2\left(\sigma_{ji}/\left(\lambda_{ji} + \lambda_{ji}^{'}\omega_{1ji}\right)\right)$, where $\omega \neq \omega_{1ji}$ and $\lambda_{ji} \neq \lambda'_{ji}$, that identifies the influence of the inner friction in the inclusion in the Voigt model. If the inclusion has a structure of the l-th rank then according to Ref. [10] and (2) the system (3) can be rewritten as follows:

$$\begin{aligned}
&\frac{\left(k_{1jil}^2 - k_{1j}^2\right)}{2\pi}\iint\limits_{Scl}\varphi_l(M)G_{Sp,j}\left(M,M^0\right)d\tau_M + \frac{\sigma_{ja}}{\sigma_{jil}}\varphi_{l-1}^0\left(M^0\right) - \\
&-\frac{\left(\sigma_{ja}-\sigma_{jil}\right)}{\sigma_{jil}2\pi}\oint\limits_{cl}G_{Sp,j}\frac{\partial\varphi_l}{\partial n}dc = \varphi_l\left(M^0\right), M^0 \in S_{cl} \\
&\frac{\sigma_{jil}\left(k_{1jil}^2 - k_{1j}^2\right)}{\sigma\left(M^0\right)2\pi}\iint\limits_{Scl}\varphi_l(M)G_{Sp,j}\left(M,M^0\right)d\tau_M + \varphi_{l-1}^0\left(M^0\right) - \\
&-\frac{\left(\sigma_{ja}-\sigma_{jil}\right)}{\sigma\left(M^0\right)2\pi}\oint\limits_{cl}G_{Sp,j}\frac{\partial\varphi_l}{\partial n}dc = \varphi_l\left(M^0\right), M^0 \notin S_{cl},
\end{aligned} \tag{4}$$

where $G_{Sp,j}(M,M^0)$ – source function of the seismic field, it coincides with the function of the expressions (3), $k_{1ji}^2 = \omega^2\left(\sigma_{ji}/\left(\lambda_{ji} + \lambda_{ji}^{'}\,\omega_{1ji}\right)\right)$ – wave number for the longitudinal wave, *ji* and *ja* coincides with expressions from (3), $l = 1\ldots L$ – number of the hierarchic level, $\underset{l}{\vec{u}} = \text{grad}\varphi_l$, φ_l^0 – potential of the normal seismic field in the layered medium by absent of heterogeneity of previous rank, if $l = 2\ldots L$ $\varphi_l^0 = \varphi_{l-1}$, if $l = 1$, $\varphi_l^0 = \varphi^0$, that coincides with the corresponding case (3).

If by transition on the next hierarchic level the axis of two-dimensionality does not change and only the geometry of the sections of included structures change, then similarly (4) we can develop the iteration process of seismic field modeling (the case of distribution only longitudinal wave). The iteration process relates to modeling of displacement vector by transition from previous (rank) level to a next (rank). Inside each hierarchic rank the integral-differential equation and the integral-differential expression is calculated using the algorithm (4). If on some hierarchical rank the structure of the local heterogeneity divides on some heterogeneities, then the integrals in the expression (4) are calculated over all heterogeneities. In that algorithm we consider the case when the physical properties for one and the same rank are equal, the boundaries cab be only different.

It must be noted, that the structure of the integral-differential equations remain the same also for the case of elastic inclusion, but the vector $\vec{u_l} = \text{grad}\varphi_l$ now depends from two additional parameters for each rank: $\lambda'_{jil}, \omega_{1jil}$, that can lead the system to a resonant state.

2.2. Modeling of elastic transversal wave diffraction on 2D plastic hierarchic heterogeneity located in the *N*-layered medium

Similarly to (4) it can be developed the same process for modeling an elastic transversal wave in *N*-layered medium with 2D hierarchic structure with arbitrary section morphology with use the integral equations from the paper [10] and (2), where $\mu_{jil} = \mu_{jil}^e + \mu_{jil}'\,\omega_{2jil}$.

$$
\begin{aligned}
&\frac{\left(k_{2jil}^2 - k_{2j}^2\right)}{2\pi} \iint\limits_{SCl_l} u_{xl}(M) G_{Ss,j}\left(M, M^0\right) d\tau_M + \frac{\mu_{ja}}{\mu_{jil}} u_{x(l-1)}^0\left(M^0\right) + \\
&+ \frac{\left(\mu_{ja} - \mu_{jil}\right)}{\mu_{jil} 2\pi} \oint\limits_{Cl} u_{xl}(M) \frac{\partial G_{Ss,j}}{\partial n} dc = u_{xl}\left(M^0\right), M^0 \in S_{Cl} \\
&\frac{\mu_{jil}\left(k_{2jil}^2 - k_{2j}^2\right)}{\mu\left(M^0\right) 2\pi} \iint\limits_{SCl_l} u_{xl}(M) G_{Ss,j}\left(M, M^0\right) d\tau_M + u_{x(l-1)}^0\left(M^0\right) + \\
&+ \frac{\left(\mu_{ja} - \mu_{jil}\right)}{\mu\left(M^0\right) 2\pi} \oint\limits_{Cl} u_{xl}(M) \frac{\partial G_{Ss,j}}{\partial n} dc = u_{xl}\left(M^0\right), M^0 \in S_{Cl}
\end{aligned}
\tag{5}
$$

where $G_{Ss,j}(M,M^0)$ – source function of the seismic field for the considering problem, it coincides with the Green function formulated in the paper [10] for the appropriate problem, $k_{2jil}^2 = \omega^2\left(\sigma_{jil}/\mu_{jil}\right)$ – wave number of the transversal wave, $\mu_{jil} = \mu_{jil}^e + \mu'_{jil}\,\omega_{2jil}$, μ – constant of Lame, u_{xl} – component of the displacement vector of the rank l,, l = 1…L – number of the hierarchic rank, u_{xl}^0 – component of the displacement vector of the previous rank in the layered medium when the heterogeneity of the previous rank is absent, if l = 2…L, $u_{xl}^0 = u_{x(l-1)}$, if l = 1, $u_{xl}^0 = u_x^0$, that coincides with the corresponding expression for the normal field in the paper [10].

Thus the iteration processes (4) and (5) allow to define by given modules of elasticity inside the layered medium that include the hierarchic heterogeneity with additional plastic parameters which depend from the frequency of inner oscillations of the medium inclusion the space frequency distribution of the components of the acoustic field on each hierarchic level. Then, using the known formula of the book [11], for each hierarchic level we can define the distribution of the components of the deformation tensor and stress tensor using the distribution of the components of the displacement vector, that depends not only from the influenced frequency, but from the frequency that is defined by the inner friction. On each hierarchic level it can be itself. Interacting with the influenced frequency medium creep state or resonant excitations can be occurred. That information plays a significant role for estimation of medium state, depending from the hierarchic structure and degree of its change.

2.3. Modeling sound diffraction on 2D anomaly dense hierarchical heterogeneity, located in a *N*-layered elastic medium

In the paper [6] it was described the algorithm of modeling of sound diffraction on elastic hierarchic inclusion, located in the *J*-th layer of the *N*-layered medium. $G_{Sp,j}(M,M^0)$ – is a source function of the seismic field, for which is formulated the boundary problem in the paper [6], $k_{1ji}^2 = \omega^2\left(\sigma_{ji}/\lambda_{ji}\right)$ – wave number for the longitudinal wave, indexes *ji* and *ja*, λ, σ, ω, $\vec{u} = grad\varphi$, φ^0, $\varphi_{ji}^0 = \varphi_{ja}^0$ are the same, that are described in the paragraph 1. Let us consider that the elastic parameters of the hierarchic inclusion for all ranks and the layer, where it is located are equal, but the density of the hierarchic inclusion for all ranks differs from the density of the layer where the inclusion is located. Then the system of equations [6] can be rewritten as follows:

$$\begin{aligned}
&\frac{\left(k_{1ji}^2-k_{1j}^2\right)}{2\pi}\iint\limits_{Scl}\varphi_l(M)G_{Sp,j}\left(M,M^0\right)d\tau_M+\frac{\sigma_{ja}}{\sigma_{ji}}\varphi_{l-1}^0\left(M^0\right)-\\
&-\frac{\left(\sigma_{ja}-\sigma_{jil}\right)}{\sigma_{jil}2\pi}\oint\limits_{Cl}G_{Sp,j}\frac{\partial\varphi_l}{\partial n}dc=\varphi\left(M^0\right),M^0\in S_{Cl}\\
&\frac{\sigma_{jil}\left(k_{1jil}^2-k_{1j}^2\right)}{\sigma\left(M^0\right)2\pi}\iint\limits_{Scl}\varphi_l(M)G_{Sp,j}\left(M,M^0\right)d\tau_M+\varphi_{l-1}^0\left(M^0\right)-\\
&-\frac{\left(\sigma_{ja}-\sigma_{jil}\right)}{\sigma\left(M^0\right)2\pi}\oint\limits_{Cl}G_{Sp,j}\frac{\partial\varphi_l}{\partial n}dc=\varphi_l\left(M^0\right),M^0\notin S_{Cl},
\end{aligned}\tag{6}$$

where $G_{Sp,j}(M,M^0)$ – a source function of seismic field, the boundary problem of it is formulated in the paper [6], $k_{1jil}^2=\omega^2\left(\sigma_{jil}/\left(\lambda_{jil}\right)\right.$; $\lambda_{jil}=\lambda_{ja}$; –wave number for the longitudinal wave, $l = 1\ldots L$ – number of the hierarchical level, φ_l^0 – potential of the acoustic field in the layered medium when the inclusion of the previous rank is absent, if $l = 2\ldots L$, $\varphi_l^0=\varphi_{l-1}$, if $l = 1$, $\varphi_l^0=\varphi^0$, that coincides with the corresponding expression (3).

If by transition on the next hierarchic level the axis of two-dimensionality does not change and only the geometry of the sections of included structures change, then similarly (4) we can develop the iteration process of seismic field modeling (the case of distribution only longitudinal wave). The iteration process relates to modeling of displacement vector by transition from previous (rank) level to a next (rank). Inside each hierarchic rank the integral-differential equation and the integral-differential expression is calculated using the algorithm (6). If on some hierarchical rank the structure of the local heterogeneity divides on some heterogeneities, then the integrals in the expression (6) are calculated over all heterogeneities. In that algorithm we consider the case when the physical properties for one and the same rank are equal; the boundaries can be only different.

2.4. Modeling of elastic transversal wave diffraction on 2D anomaly dense hierarchical heterogeneity, located in a *N*-layered elastic medium

Similarly to (6) it can be developed the same process for modeling distribution elastic transversal wave in N-layered medium with 2D hierarchic structure with arbitrary section morphology with use the integral equations from the paper [6].

$$\begin{aligned}
&\frac{\left(k_{2jil}^2-k_{2j}^2\right)}{2\pi}\iint\limits_{SCl_l}u_{xl}(M)G_{Ss,j}\left(M,M^0\right)d\tau_M+u_{x(l-1)}^0\left(M^0\right)=u_{xl}\left(M^0\right),M^0\in S_{Cl}\\
&\frac{\mu_{jil}\left(k_{2jil}^2-k_{2j}^2\right)}{\mu\left(M^0\right)2\pi}\iint\limits_{SCl_l}u_{xl}(M)G_{Ss,j}\left(M,M^0\right)d\tau_M+u_{x(l-1)}^0\left(M^0\right)=u_{xl}\left(M^0\right),M^0\notin S_{Cl}
\end{aligned}\tag{7}$$

where $G_{Ss,j}(M,M^0)$ – source function of the seismic field for the considering problem, it coincides with the Green function formulated in the paper [6] for the appropriate problem, $k_{2jil}^2 = \omega^2(\sigma_{jil} / \mu_{jil})$ – wave number of the transversal wave, $\mu_{jil}=\mu_{ja}$, μ – constant of Lame, u_{xl} – component of the displacement vector of the rank l, $l = 1...L$ – number of the hierarchic rank, u_{xl}^0 – component of the displacement vector of the previous rank in the layered medium when the heterogeneity of the previous rank is absent, if $l = 2...L$, $u_{xl}^0 = u_{x(l-1)}$, if $l = 1$, $u_{xl}^0 = u_x^0$, that coincides with the corresponding expression for the normal field in the paper [10].

It must be noted that the structure of the system (6) coincides with the general case, when the hierarchical heterogeneity had not only density parameters that differ from the density of its included layer, but also the elastic parameters for all ranks differ from the elastic parameters of the included layer. The difference of the system (6) consists only in the values of the wave number. Thus more sensitive to the hierarchical inclusions of anomaly density in the massif is the medium response, linked with the longitudinal wave that is also sensitive to the form of the hierarchical inclusion, than the transversal wave. That must be taken into account by mapping and monitoring such complicated geological medium.

2.5. Modeling sound diffraction on 2D anomaly stressed hierarchical heterogeneity, located in an N-layered elastic medium

In the paper [6] was described the algorithm of modeling of sound diffraction on elastic hierarchic inclusion, located in the J-th layer of the N-layered medium. $G_{Sp,j}(M,M^0)$ – source function of the seismic field, for which is formulated the boundary problem in the paper [6], $k_{1ji}^2 = \omega^2(\sigma_{ji} / \lambda_{ji})$ – wave number for the longitudinal wave, indexes ji and ja, λ, σ, ω, $\vec{u} = grad\varphi$, φ^0, $\varphi_{ji}^0 = \varphi_{ja}^0$ are the same, that are described in the paragraph 1. Let us consider that the density parameters of the hierarchic inclusion for all ranks and the layer, where it is located are equal, but the elastic parameters in the hierarchic inclusion for all ranks differ from the elastic parameters of the layer where the inclusion is located. Then the system of equations [6] can be rewritten as follows:

$$\begin{aligned} &\frac{\left(k_{1jil}^2 - k_{1j}^2\right)}{2\pi} \iint\limits_{Scl} \varphi_l(M) G_{Sp,j}(M,M^0) d\tau_M + \varphi_{l-1}^0(M^0) = \varphi_l(M^0), M^0 \in S_{Cl} \\ &\frac{\sigma_{jil}\left(k_{1jil}^2 - k_{1j}^2\right)}{\sigma(M^0) 2\pi} \iint\limits_{Scl} \varphi_l(M) G_{Sp,j}(M,M^0) d\tau_M + \varphi_{l-1}^0(M^0) = \varphi_l(M^0), M^0 \notin S_{Cl} \end{aligned} \quad (8)$$

The designations in (8) are the same as for the system of equations (6).

2.6. Modeling of elastic transversal wave diffraction on 2D anomaly stressed hierarchical heterogeneity, located in an N-layered elastic medium

Similarly to (4) it can be developed the same process for modeling distribution elastic transversal wave in N-layered medium with 2D hierarchic structure with arbitrary section morphology with use the integral equations from the paper [6].

$$\begin{gathered}
\frac{\left(k_{2jil}^2 - k_{2j}^2\right)}{2\pi} \iint\limits_{SCl_l} u_{xl}(M) G_{Ss,j}(M, M^0) d\tau_M + \frac{\mu_{ja}}{\mu_{jil}} u_{x(l-1)}^0 (M^0) + \\
+ \frac{\left(\mu_{ja} - \mu_{jil}\right)}{\mu_{jil} 2\pi} \oint\limits_{Cl} u_{x1}(M) \frac{\partial G_{Ss,j}}{\partial n} dc = u_{xl}(M^0), M^0 \in S_{Cl} \\
\frac{\mu_{jil}\left(k_{2jil}^2 - k_{2j}^2\right)}{\mu(M^0) 2\pi} \iint\limits_{SCl_l} u_{xl}(M) G_{Ss,j}(M, M^0) d\tau_M + u_{x(l-1)}^0 (M^0) + \\
+ \frac{\left(\mu_{ja} - \mu_{jil}\right)}{\mu(M^0) 2\pi} \oint\limits_{Cl} u_{x1}(M) \frac{\partial G_{Ss,j}}{\partial n} dc = u_{xl}(M^0), M^0 \in S_{Cl}
\end{gathered} \tag{9}$$

where $G_{Ss,j}(M,M^0)$ – source function of the seismic field for the considering problem, it coincides with the Green function formulated in the paper [6] for the appropriate problem, $k_{2jil}^2 = \omega^2\left(\sigma_{jil}/\mu_{jil}\right)$ – wave number of the transversal wave, $\mu_{jil} \neq \mu_{ja}$, $\sigma_{jil} = \sigma_{ja}$, μ – constant of Lame, u_{xl} – component of the displacement vector of the rank l,, $l = 1 \ldots L$ – number of the hierarchic rank, u_{xl}^0 – component of the displacement vector of the previous rank in the layered medium when the heterogeneity of the previous rank is absent, if $l = 2 \ldots L$, $u_{xl}^0 = u_{x(l-1)}$, if $l = 1$, $u_{xl}^0 = u_x^0$, that coincides with the corresponding expression for the normal field in the paper [6].

It must be noted that the structure of the system (9) coincides with the general case, when the hierarchical heterogeneity had not only density parameters that differ from the density of its included layer, but also the elastic parameters for all ranks differ from the elastic parameters of the included layer. The difference of the system (9) consists only in the values of the wave number. Thus more sensitive to the hierarchical inclusions of anomaly density in the massif is the medium response, linked with the transversal wave. That must be taken into account by mapping and monitoring such complicated geological medium.

2.7. Algorithm of modeling the distribution of the longitudinal wave in the layered medium with fluid saturated hierarchic inclusions

The idea, written in the paper [6] for solution of the direct problem for 2D case of longitudinal wave distribution through the local elastic heterogeneity with hierarchic structure, located in the *J*-the layer of N-layered medium, let us spread on the case when on the *L*-th hierarchic layer a fluid saturated porous inclusion will occur. $G_{Sp,i}(M,M^0)$ – source function of the seismic field, it coincides with the function from the paper [6]. $k_{1jil}^2 = \omega^2\left(\sigma_{jil}/\lambda_{jil}\right)$ – wave number for the longitudinal wave. Indexes *ji* and *ja*, λ, σ, ω, $\vec{u} = grad\ \varphi$, φ^0, $\varphi_{ji}^0 = \varphi_{ja}^0$ are the same, that are described in the paragraph 1. $l = 1 \ldots L\text{-}1$ – number of the hierarchic level.

If by transition on the next hierarchic level the axis of two-dimensionality does not change and only the geometry of the sections of included structures change, then similarly (4) we can develop the iteration process of seismic field modeling (the case of distribution only longitudinal wave). The iteration process relates to modeling of displacement vector by transition from previous (rank) level to a next (rank). Inside each hierarchic rank the integral-differential equation and the integral-differential expression is calculated using the algorithm (4). If on some hierarchical rank the structure of the local heterogeneity divides on some heterogeneities, then the integrals in the expression (4) are calculated over all heterogeneities. In that algorithm we consider the case when the physical properties for one and the same rank are equal; the boundaries can be only different. If l = L, inside these hierarchic level the porous fluid saturated inclusion occurred. Then the system (4) with account [12] will be rewritten:

$$\begin{gathered}
\frac{\left(k_{1jil}^2 - k_{1j}^2\right)}{2\pi} \iint\limits_{Scl} \varphi_l(M) G_{Sp,j}\left(M, M^0\right) d\tau_M + \frac{\sigma_{ja}}{\sigma_{jil}} \varphi_{l-1}^0\left(M^0\right) - \\
-\frac{\left(\sigma_{ja} - \sigma_{jil}\right)}{\sigma_{jil} 2\pi} \oint\limits_{cl} G_{Sp,j} \frac{\partial \varphi_l}{\partial n} dc = \left(\varphi_l\left(M^0\right) + \alpha p_2\right), M^0 \in S_{cl} \\
\frac{\sigma_{jil}\left(k_{1jil}^2 - k_{1j}^2\right)}{\sigma\left(M^0\right) 2\pi} \iint\limits_{Scl} \varphi_l(M) G_{Sp,j}\left(M, M^0\right) d\tau_M + \varphi_{l-1}^0\left(M^0\right) - \\
-\frac{\left(\sigma_{ja} - \sigma_{jil}\right)}{\sigma\left(M^0\right) 2\pi} \oint\limits_{cl} G_{Sp,j} \frac{\partial \varphi_l}{\partial n} dc = \varphi_l\left(M^0\right), M^0 \notin S_{cl},
\end{gathered} \tag{9a}$$

where $\alpha = 1 - \chi - \frac{K}{K_0}$, $K=\lambda$ – module of all-around compression, χ – porosity, K_0 – true modulus of phase compressibility, pore hydrostatic pressure p_2. If $l = L + 1$ and on the next level the hierarchic heterogeneity is again elastic, for continuing the iteration process we can use again the equations (4).

3. Defining of the 2-D surface of the anomaly stressed hierarchical object, located in the layered blocked geological medium using the data of acoustic monitoring

In Ref. [13], the concept of a step-by-step interpretation of a variable electromagnetic field was proposed. At the first stage, the parameters of the normal section, or the parameters of the enclosing one-dimensional non-magnetic medium, are anomalous conductive or magnetic inclusions. In the second stage, an anomalous alternating electromagnetic field is selected by a system of singular sources placed in a horizontally layered medium with geoelectrical parameters determined at the first stage. At the third stage, the theoretical inverse problem is solved, i.e. At given geoelectrical parameters of the host environment for a set of parameters of non-homogeneities, the contours of this heterogeneity are determined. We obtain explicit integral-differential equations of the theoretical inverse scattering problem for two-dimensional and three-dimensional alternating and three-dimensional stationary electromagnetic fields in the framework of models: a conductive or a magnetic body in the ν-th layer of a conductive N-layered half-space.

In this chapter, using the approach presented [14, 15], an algorithm is developed for obtaining the equation of the theoretical inverse problem for an acoustic field (transverse acoustic wave) for the elastic anomaly stressed hierarchical heterogeneity of the k-th rank, whose density coincides with the density of the host medium for all hierarchical ranks, located in the ν-th layer of the elastic N-layered half-space.

Let a simply connected domain D of the Euclidean space R^2, bounded by a continuously differentiable closed curve is located in the ν-th layer of the N-layered half-space. Suppose that this domain contains inside K simply connected hierarchical inclusions, bounded by continuously differentiable closed curves ∂D_k and extending parallel to the axis OX. The boundaries l_i of the layers P_j (j = 1, ..., N) are parallel to the OY axis of the XOY plane of the Cartesian coordinate system. The axis OZ is directed vertically downward. We place the origin of the coordinate system on the upper boundary of the surface of the first layer and match it with the point that is the projection onto OY of the point, with respect to which the domain D is stellar. Let U (y, z) be complex-valued twice continuously differentiable function that satisfy the two-dimensional scalar Helmholtz equation:

$$\Delta U + c(M)U = -f(M), \tag{10}$$

where $\Delta = \frac{\partial^2}{\partial y^2} + \frac{\partial^2}{\partial z^2}$;

$$c(M) = \begin{cases} c_j; M \in \Pi_j/\overline{D}\,(j = 0, \ldots n) \\ c_{ak}; M \in D_k(k = 1, \ldots K) \end{cases}. \tag{11}$$

Let the function $U^1(y, z)$ satisfies the equation:

$$\Delta U^1 + p(M)U^1 = -f(M), \tag{12}$$

$$p(M) = \begin{cases} c_j; M \in \Pi_j/\overline{D}\,(j = 0, \ldots n) \\ c_\nu; M \in D_k(k = 1, \ldots K) \end{cases}. \tag{13}$$

Let us first consider the case, when $k = 1$. For $M \in R^2\ \overline{D}(j = 0, \ldots N)$.We shall define:

$$U^+(M) = U(M) - U^1(M) \tag{14}$$

Function $U^+(M)$ satisfies the Eq. (10). On the boundaries l_j of the layers P_j the following boundary conditions are fulfilled:

$$\begin{aligned} &U_j = U_{j+1}; \\ &U_j^+ = U_{j+1}^+; M \in l_j(j = 1, \ldots, n-1); \\ &U_j^1 = U_{j+1}^1; \end{aligned} \tag{15}$$

$$b_j \frac{\partial U_j}{\partial n} = b_{j+1} \frac{\partial U_{j+1}}{\partial n}; b_j \frac{\partial U_j^+}{\partial n} = b_{j+1} \frac{\partial U_{j+1}^+}{\partial n}; b_j \frac{\partial U_j^1}{\partial n} = b_{j+1} \frac{\partial U_{j+1}^1}{\partial n}; M \in l_j, \tag{16}$$

b_j are complex coefficients $(j = 0, \ldots, N)$ and in general case:$b_j \neq b_{j+1}$; on the contour ∂D_k: for $k = 1$:

$$U_\nu = U_\nu^+ + U_\nu^1. \tag{17}$$

Function U_ν satisfies the equation:

$$\Delta U_v + c_v(M)U_v = -f(M). \tag{18}$$

U_v^+ – is a function U^+ in the layer $P_v \notin D$; U_v^1 – is a function U^1 in the layer $P_v \notin D$; in the domain D for $k = 1$:

$$U_a = U_a^+ + U_a^1; M \in \overline{D}; \Delta U_a + c_a U_a = 0. \tag{19}$$

Boundary conditions on ∂D ($k = 1$):

$$U_a^+ = U_v^+, \quad b_a \frac{\partial U_a}{\partial n} - b_v \left(\frac{\partial U_v^+}{\partial n} + \frac{\partial U_v^1}{\partial n} \right) = 0, \tag{20}$$

By $M \to \infty$ the functions $U(M), U^+(M), U^1(M)$ satisfy the radiation condition [16]. The algorithm of calculation of function U^1 for the electromagnetic case is written [13].

Let us introduce the function $G(M, M_0)$, that satisfies the following equation:

$$\Delta G + p(M)G = -\delta(M, M_0), \tag{21}$$

and the boundary conditions (15, 16), by $M \to \infty$ the function G satisfies the radiation condition [16], by $M \to M_0$ the function G has a singularity as: $\ln 1/\rho(M, M_0)$:

$$\rho(M, M_0) = \sqrt{(y - y_0)^2 + (z - z_0)^2}. \tag{22}$$

Algorithm of calculation of function G for the case, when the domain D is located in the v – th layer, is written [13]. Let us introduce the function G^a, that coincides with the fundamental solution of the Eq. (11) for $k = 1$. Let us use the Green formula [16] for two functions $U^+, G; \left(M \in R^2 \backslash \overline{D}, M_0 \in P_i\right)$ in each layer P_j(j=0, …, N). Let us fulfill the procedure similarly [17]: let us multiply the defined expressions for each layer on b_j reciprocally, j=0, … , N and add them term by term with account (11–14), (16) and (17). As a result we receive:

$$2\pi U^+(M_0) = -(b_v/b_i) \int_{\partial D} \left(U_v^+(M) \frac{\partial G(M, M_0)}{\partial n} - G(M, M_0) \frac{\partial U_v^+}{\partial n} \right); M \in \Pi_v; M_0 \in \Pi_i. \tag{23}$$

In the domain D let us use the Greens formula for the two functions $U_a(M), G^a(M, M_0)$. As a result we receive:

$$0 = \int_{\partial D} \left(U_a(M) \frac{\partial G^a(M, M_0)}{\partial n} - G^a(M, M_0) \frac{\partial U_a}{\partial n} \right) dl. \tag{24}$$

Let us add expressions (23) and (24), taking into account (20) and (21), and also the expression (13):

$$0 = \left(-{}^{b_v}/_{b_i} \right) \int_{\partial D} \left(U_v^1(M) \frac{\partial G(M, M_0)}{\partial n} - G(M, M_0) \frac{\partial U_v^1}{\partial n} \right) dl; M \in \overline{D}; M_0 \in \Pi_i. \tag{25}$$

Then we shall receive:

$$2\pi U^{+}(M_0) = \int\limits_{\partial D} \left(\left(U_{\nu}^{+}(M) + U_{\nu}^{1}(M)\right) \left(\frac{\partial G^{a}(M, M_0)}{\partial n} - \left({}^{b_\nu}/_{b_i} \right) \frac{\partial G(M, M_0)}{\partial n} \right) - \right.$$
$$\left. - b_\nu \left(\frac{\partial U_{\nu}^{+}}{\partial n} + \frac{\partial U_{\nu}^{1}}{\partial n} \right) \left(\left({}^{1}/_{b_a} \right) G^{a}(M, M_0) - \left({}^{1}/_{b_i} \right) G(M, M_0) \right) \right) dl. \tag{26}$$

Eq. (26) is the explicit equation of the theoretical inverse problem for the two-dimensional scalar Helmholtz equation in a layered medium with homogeneous inclusion for given values of the boundary conditions [13–15]. As a result of the solution of the integral-differential equation (26) related to the function $r\ (\varphi)$, that describes the contour of the sought homogeneous object, it is possible to determine it for known values of the physical parameters of the host medium and the desired object, and also for given values of the functions.

According to [5, 8], the problem of diffraction of a linear polarized elastic transverse wave by a two-dimensional elastic heterogeneity of a hierarchical type located in a layer of an N-layer medium within the framework of the described model reduces to solving a similar problem with the following changes. The equation of the theoretical inverse problem (26) for the scalar Helmholtz equation, to which our problem reduces, remains valid by that:

$$b_\nu = \xi_\nu; b_i = \xi_i; b_a = \xi_a; \tag{27}$$

$\xi_\nu, \xi_i, \xi_a, \rho_\nu, \rho_i, \rho_a$ – the values of the parameter Lame and the density in the ν -th layer, in the layer where the point M_0 is located and inside the heterogeneity at k = 1. An important difference between the present problem and the one considered above is that ρ_a=ρ_ν for all k, physically, this means that the anomaly in the acoustic field is created by an anomaly of the stressed state of the medium and can be associated with a focus of either a rock shock or earthquakes.

$$U^{+} = u_x^{+}; U_\nu^{+} = u_{x\nu}^{+}; U_\nu^{1} = u_{x\nu}^{1}, \tag{28}$$

where u_x is the component of the displacement vector, different from zero for the selected model.

$$G(M, M_0) = G_{SS}(M, M_0); G^{a}(M, M) = G_{SS}^{a}(M, M_0); \partial D, dl,$$
$$k_{2a}^{2} = \omega^2 \frac{\rho_\nu}{\xi_a}; k_{2\nu}^{2} = \omega^2 \frac{\rho_\nu}{\xi_\nu}. \tag{29}$$

The algorithm for calculating the Green's function $G_{SS}(M,M_0)$ was written in Ref. [18]. Thus, the equation of the theoretical inverse problem for $k = 1$ is written in the form:

$$2\pi u_x^{+}(M_0) = \int\limits_{\partial D1} \left(\left(u_{x\nu}^{+}(M) + u_{x\nu}^{1}(M)\right) \left(\frac{\partial G_{SS}^{a1}(M, M_0)}{\partial n} - \left({}^{\xi_\nu}/_{\xi_i} \right) \frac{\partial G_{SS}(M, M_0)}{\partial n} \right) - \right.$$
$$\left. - \xi_\nu \left(\frac{\partial u_x^{+}}{\partial n} + \frac{\partial u_{x\nu}^{1}}{\partial n} \right) \left(\left({}^{1}/_{\xi_{a1}} \right) G_{SS}^{a1}(M, M_0) - \left({}^{1}/_{\xi_i} \right) G_{SS}(M, M_0) \right) \right) dl. \tag{30}$$

Let $k = 2$, that is the sought object is a hierarchic inclusion with elastic parameter Lame ξ_1 and density ρ_v, ∂D_1 – contour of the external inclusion and with elastic parameter Lame ξ_2, density ρ_v, ∂D_2 – contour of the inner inclusion. The inclusions are uncoordinated. It is needed to define the two contours. For solution of our problem in the expression (27) we substitute ξ_a=ξ_1, and in the expression (29):

$$\partial D = \partial D_1; dl = dl_1; G^a_{SS} = G^{a1}_{SS}; \tfrac{\partial}{\partial n}\left(G^a_{SS}\right) = \tfrac{\partial}{\partial n}\left(G^{a1}_{SS}\right); k^2_{2a} = \tfrac{2a12}{k} = \omega^2 \tfrac{\rho_v}{\xi_{a1}}.$$

Solving Eq. (30) related to the function $r_1(\varphi)$, that describes the contour ∂D_1, we calculate the functions: $u_x; u^+_x; u^1_x$ by the algorithm for solving the direct problem (9), (31) and (32) inside and outside the heterogeneity placed in a layered medium, u^1_x the elastic field component in a layered medium in the absence of a heterogeneity.

$$\begin{aligned} u_x(M_0) &= \frac{\xi_v}{\xi_a} u^1_x(M_0) + \frac{k^2_{2a} - k^2_{2v}}{2\pi} \iint\limits_{S1} u_x(M) G_{SS}(M, M_0) dS_1 + \\ &+ \frac{\xi_v - \xi_a}{2\pi\, \xi_a} \int\limits_{\partial D_1} u_x(M) \frac{\partial G_{SS}}{\partial n} dl; M_0 \in S_1, \end{aligned} \tag{31}$$

$$\begin{aligned} u_x(M_0) &= u^1_x(M_0) + \frac{\xi_a\left(k^2_{2a} - k^2_{2v}\right)}{2\pi\, \xi(M_0)} \iint\limits_{S1} u_x(M) G_{SS}(M, M_0) dS_1 + \\ &+ \frac{(\xi_v - \xi_a)}{2\pi\, \xi(M_0)} \int\limits_{\partial D_1} u_x(M) \frac{\partial G_{SS}}{\partial n} dl; M_0 \notin S_1. \end{aligned} \tag{32}$$

This completes the first iteration cycle, and we proceed to the second iterative cycle $k = 2$. The calculated function $u_x(M_0)$ (32) is denoted as: $u_x^{1(k-1)}$ (33), in the expression (27) ξ_a=ξ_2, in the expression (29):

$$\partial D = \partial D_2; dl = dl_2; G^a_{SS} = G^{a2}_{SS}; \frac{\partial}{\partial n}\left(G^a_{SS}\right) = \frac{\partial}{\partial n}\left(G^{a2}_{SS}\right); k^2_{2a} = k^2_{2a2} = \omega^2 \frac{\rho_v}{\xi_{a2}}. \tag{33}$$

The Eq. (30) can be rewritten as following:

$$\begin{aligned} 2\pi u^+_x(M_0) &= \int\limits_{\partial D2} \left(\left(u^+_{xv}(M) + u^{1(k-1)}_{xv}(M)\right)\left(\frac{\partial G^a_{SS}(M, M_0)}{\partial n} - \left({}^{\xi_v}/_{\xi_i}\right)\frac{\partial G_{SS}(M, M_0)}{\partial n}\right) - \right. \\ &- \xi_v \left(\frac{\partial u^+_x}{\partial n} + \frac{\partial u^{1(k-1)}_{xv}}{\partial n}\right)\left(\left({}^1/_{\xi_{a2}}\right) G^a_{SS}(M, M_0) - \left({}^1/_{\xi_i}\right) G_{SS}(M, M_0)\right)\Big) dl_2; \end{aligned} \tag{34}$$

We solve the Eq. (34) relatively the function $r_2(\varphi)$ that describes the contour ∂D_2.If $K = 2$, then the problem is solved, if $k > 2$, $k = k + 1$, the iteration process is continued.

We solve the functions:

$$u^{k-1}_x; u^{+(k-1)}_x, \tag{35}$$

using the algorithm of the direct problem inside and outside the hierarchical heterogeneity of the rank $k - 1$, located into the layered medium (the physical parameters of the layered medium are not changed) (9), (36) and (37).

$$u_x^{k-1}(M_0) = \frac{\xi_v}{\xi_{a(k-1)}} u_x^{1(k-2)}(M_0) + \frac{k_{2a(k-1)}^2 - k_{2v}^2}{2\pi} \iint\limits_{S(k-1)} u_x^{k-1}(M) G_{SS}(M, M_0) dS_{(k-1)} + \\ + \frac{\xi_v - \xi_{a(k-1)}}{2\pi \xi_{a(k-1)}} \int\limits_{\partial D(k-1)_1} u_x^{k-1}(M) \frac{\partial G_{SS}}{\partial n} dl_{k-1}; M_0 \in S_{(k-1)}, \tag{36}$$

$$u_x^{k-1}(M_0) = u_x^{1(k-2)}(M_0) + \frac{\xi_a (k_{2a}^2 - k_{2v}^2)}{2\pi\, \xi(M_0)} \iint\limits_{S(k-1)} u_x^{k-1}(M) G_{SS}(M, M_0) dS_1 + \\ + \frac{(\xi_v - \xi_a)}{2\pi\, \xi(M_0)} \int\limits_{\partial D(k-1)_1} u_x^{k-1}(M) \frac{\partial G_{SS}}{\partial n} dl_{k-1}; M_0 \notin S_{(k-1)}. \tag{37}$$

The calculated function $u_x^{k-1}(M_0)$ (28) we denote as:

$$u_x^{1(k-1)} \tag{38}$$

In the expression (27) $\xi_a = \xi_k$, in the expression (29):

$\partial D = \partial D_k; dl = dl_k; G_{SS}^a = G_{SS}^{ak}; \frac{\partial}{\partial n}(G_{SS}^a) = \frac{\partial}{\partial n}(G_{SS}^{ak}); k_{2a}^2 = k_{2ak}^2 = \omega^2 \frac{\rho_v}{\xi_{ak}}$..

Eq. (30) is rewritten as following:

$$2\pi u_x^+(M_0) = \int\limits_{\partial D} \left(\left(u_{xv}^+(M) + u_{xv}^{1(k-1)}(M) \right) \left(\frac{\partial G_{SS}^a(M, M_0)}{\partial n} - \left({}^{\xi_v}/_{\xi_i} \right) \frac{\partial G_{SS}(M, M_0)}{\partial n} \right) - \\ - \xi_v \left(\frac{\partial u_x^+}{\partial n} + \frac{\partial u_{xv}^{1(k-1)}}{\partial n} \right) \left(({}^1/_{\xi_a}) G_{SS}^a(M, M_0) - ({}^1/_{\xi_i}) G_{SS}(M, M_0) \right) \right) dl_2; \tag{39}$$

We solve the Eq. (39) relatively the function $r_k(\varphi)$ that describes the contour ∂D. $k = k + 1$. Iteration process (35–39) continues up to $k = K$.

4. Conclusions

The chapter considers the problem of modeling a seismic field acoustic approximation in a layered medium with inclusions of a hierarchical structure. Algorithms of modeling in the seismic case in the acoustic approximation for 2-D plastic heterogeneity are constructed.

Comparing expressions (6) and (7), (8) and (9), we can draw the following conclusions. When constructing a complex seismic gravity model without taking into account the anomalous effect of a stress–strain state inside the inclusion, an analysis of the anomalous acoustic effect using data on the propagation of a longitudinal wave shows that it is more sensitive also to the inclusion form, in comparison with the acoustic effect of the propagation of a transverse wave. However, it follows from these expressions that elastic parameters in the seismic model cannot be neglected in the massif, and they influence the interpretation of the values of the unknown anomalous densities. If these values are used in the construction of a density gravitational model, then these density values will not reflect the material composition of the analyzed medium. In the construction of an anomalously stressed geomechanical model without taking into account the anomalous effect of density heterogeneities within the inclusion, an analysis of the anomalous acoustic effect using data on the propagation of a transverse wave shows that it is more sensitive also to the inclusion form, in comparison with the acoustic effect of the propagation of a longitudinal wave. However, it follows from these expressions that the influence of density parameters in the massif in the seismic model cannot be neglected, and they influence the interpretation of the values of the unknown anomalous elastic parameters that cause the anomalous stress state. If these values are used in the construction of a geomechanical model, these values of the elastic parameters will not reflect the stress state of the analyzed medium.

The proposed simulation algorithm, the mapping and monitoring method for a heterogeneous, complex two-phase medium, can be used to control the production of viscous oil in mine conditions and light oil in sub horizontal wells. The requirements of efficient for the most economic indicators and the most complete extraction of hydrocarbons in the fields dictate the need for the creation of new geotechnology for the development of oil and gas fields based on fundamental advances in geophysics and geomechanics [19].

Additionally it is of interest with the use of the obtained algorithms to investigate the question of studying the connection between the strain and strain tensors at each hierarchical level and the possible deviation from the generalized Hooke's law.

In the third part of our chapter we considered the problem of constructing an algorithm for solving the inverse problem using the equation of the theoretical inverse problem for the 2-D Helmholtz equation. An explicit equation is obtained for the theoretical inverse problem for the scattering of a linearly polarized elastic wave in a layered elastic medium with a hierarchical elastic anomal stressed inclusion, whose density for all ranks is equal to the density of the enclosing layer. An iterative algorithm for determining the contours of non-axial inclusions of the k-th rank in a hierarchical structure with successive use of the solution of the direct problem of calculating the elastic field of k-1 rank is constructed. With the increase in the degree of hierarchy of the structure of the medium, the degree of spatial nonlinearity of the distribution of the components of the acoustic field increases, which involves the elimination of methods for linearizing the problem when creating interpretation methods. This problem is inextricably linked with the solution of the inverse problem for the propagation of the acoustic field in such complex media using explicit equations of the theoretical inverse problem. For the first time an equation for determining the surface of anomalously stressed inclusion in a hierarchical layered block environment was derived from the data of acoustic monitoring. In practice, using this algorithm, we can localize from the acoustic monitoring data the area of a potential hotbed or a forthcoming earthquake and estimate the degree of anomalous elastic stresses.

Author details

Olga Hachay[1]* and Andrey Khachay[2]

*Address all correspondence to: olgakhachay@yandex.ru

1 Institute of Geophysics UB RAS, Ekaterinburg, Russia

2 Ural Federal University, IESM, Ekaterinburg, Russia

References

[1] Khachay O. Geosynergetic: Theory, methodology, experiment. In: Spichak V, editor. Complex Analysis of Electromagnetic and Other Geophysical Data. Moscow: KRASAND; 2011. p. 155-176 (in Russian)

[2] Adushkin A, Tsvetkov V. Influence of structure and geodynamics on the stress state of the Earth's. In: . Problems of the Mechanics Rocks. St. Petersburg; 1997. p. 7-12 (in Russian)

[3] Shkuratnik V, Lavrov A. Effects of Memory in Rocks. Physical Patterns, Theoretical Models. Moscow: Publishing House of the Academy of Mining Sciences; 1997. p. 158 (in Russian)

[4] Catalog of Rock Impacts on ore and non-ore Deposits. Leningrad: VNIMI; 1986. p. 183 (in Russian)

[5] Khachay O, Khachay A. On the joining of seismic and electromagnetic active methods for mapping and monitoring the state of two-dimensional heterogeneities in an N-layer medium. Bulletin of SSSU Series of "Computer Technologies, Management, Radio Electronics". 2011;**2**:49-56 (in Russian)

[6] Khachay O, Khachay A. Modeling of the electromagnetic and seismic field in hierarchically inhomogeneous media. Bulletin SSSU Series "Computational Mathematics and Informatics". 2013;**2**:49-56 (in Russian)

[7] Khasanov M, Bulgakova G. Non-linear and No Equilibrium Effects in Rheological Complex Media. Institute for Computer Research: Moscow Izhevsk; 2003 288 p (in Russian)

[8] Kupradze V. Boundary-Value Problems in the Theory of Oscillations and Integral Equations. Moscow-Leningrad; 1950. p. 280 (in Russian)

[9] Kol'skii G. Waves of Stress in Solids. Edition of Foreign Literature: Moscow; 1955 194 p

[10] Khachay O, Khachay OY, Khachay A. New methods of Geoinformatics monitoring of wave fields in hierarchical environments. Geoinformatics. 2015;**3**:45-51 (in Russian)

[11] Lurie A. Spatial Problems in the Theory of Elasticity. Moscow: State Ed. for Technical and Theoretical Literature; 1956. p. 435 (in Russian)

[12] Frenkel Y. To the theory of seismic and seism electric phenomena in moist soil. Izvestija of the USSR Academy of Sciences, Series of Geographic and Geophysical. 1944;**8**:133-150 (in Russian)

[13] Khachay O. Mathematical modeling and interpretation of a variable electromagnetic field in the inhomogeneous crust and upper mantle of the Earth Doctoral thesis. IGF UB RAS: Sverdlovsk; 1994. p. 314 (in Russian)

[14] Khachay O. On the interpretation of two-dimensional variables and three-dimensional stationary anomalies of the electromagnetic field, izv.AN.USSR. Physics of the Earth. 1989;**10**:50-58 (in Russian)

[15] Khachay O. On the solution of the inverse problem for three-dimensional variable electromagnetic fields, izv.AN.USSR. Physics of the Earth. 1990;**2**:55-59

[16] Stratton D. Theory of Electromagnetism. Moscow–Leningrad: Publishing of OGIZ; 1948. p. 409 (in Russian)

[17] Dmitriev V. Diffraction of 2-D electromagnetic field on cylindrical bodies, located in layered media. Computational Methods and Programming. Moscow: Publishing of Moscow State University. 1965. **3**. 307–315. (in Russian)

[18] Khachay A. Algorithms for mathematical modeling alternating electromagnetic and seismic fields in the source approach PhD thesis. Ekaterinburg: UrSU; 2007. p. 176 (in Russian)

[19] Oparin V, Simonov B, Yushkin V, Vostrikov V, Pogarskii Y, Nazarov L. Geomechanical and Technical Fundamentals of Enhanced oil Recovery in Vibro-Wave Technologies. Novosibirsk: Science; 2010. p. 404 (in Russian)

Section 2

Experimental Studies of Acoustic Waves

Chapter 5

In-Fiber Acousto-Optic Interaction Based on Flexural Acoustic Waves and Its Application to Fiber Modulators

Miguel Ángel Bello Jiménez,
Gustavo Ramírez-Meléndez,
Erika Hernández-Escobar, Andrés Camarillo-Avilés,
Rosa López-Estopier, Olivier Pottiez,
Cristian Cuadrado-Laborde, Antonio Díez,
José L. Cruz and Miguel V. Andrés

Additional information is available at the end of the chapter

http://dx.doi.org/10.5772/intechopen.71411

Abstract

The design and implementation of in-fiber acousto-optic (AO) devices based on acoustic flexural waves are presented. The AO interaction is demonstrated to be an efficient mechanism for the development of AO tunable filters and modulators. The implementation of tapered optical fibers is proposed to shape the spectral response of in-fiber AO devices. Experimental results demonstrate that the geometry of the tapered fiber can be regarded as an extra degree of freedom for the design of AO tunable attenuation filters (AOTAFs). In addition, with the objective of expanding the application of AOTAFs to operate as an amplitude modulator, acoustic reflection was intentionally induced. Hence, a standing acoustic wave is generated which produces an amplitude modulation at twice the acoustic frequency. As a particular case, an in-fiber AO modulator composed of a double-ended tapered fiber was reported. The fiber taper was prepared using a standard fusion and pulling technique, and it was tapered down to a fiber diameter of 70 μm. The device exhibits an amplitude modulation at 2.313 MHz, which is two times the acoustic frequency used (1.1565 MHz); a maximum modulation depth of 60%, 1.3 dB of insertion loss, and 40 nm of modulation bandwidth were obtained. These results are within the best results reported in the framework of in-fiber AO modulators.

Keywords: acousto-optic interaction, acousto-optic filter, acousto-optic modulator

1. Introduction

The acousto-optic (AO) effect based on flexural acoustic waves in optical fibers has received a great deal of attention for the development of many practical AO devices such as frequency shifters, attenuators, and tunable filters [1–3]. In recent years, novel AO devices have been proposed and developed to explore applications of in-fiber AO interaction in a different way, for example, by implementing standing flexural acoustic waves [4, 5] or by the use acoustic waves in short packets [6]. All these approaches offer the advantages of being dynamic devices with tunable amplitude and spectral responses. Its operation principle relies on the intermodal coupling induced by flexural acoustic waves along the fiber. The propagation of the acoustic field produces a periodic perturbation of the refractive index that leads to an intermodal resonant coupling between the fundamental core mode and some asymmetric cladding modes of the fiber [1–3]. This AO effect produces a similar response to that obtained with a conventional long-period grating (LPG), but in this case, the transmission response can be controlled dynamically by the amplitude and frequency of the acoustic wave.

In the framework of all-fiber modulators that exploit the dynamic properties of AO interaction, several alternatives have been proposed and demonstrated [4, 6–15]. Among them, we have proposed a modulation technique based on the intermodal coupling induced by standing flexural acoustic waves in a standard optical fiber [4, 15]. This novel type of modulator provides a stable amplitude modulation that can be used to realize mode-locking operation in a fiber laser system [4, 15]. However, its major shortcoming is the reduced modulation bandwidth (~1.5 nm), which could be a detrimental factor for systems that require broader optical bandwidths. In this chapter we report an improved version of the AO modulator by implementing tapered optical fibers as the mechanism to optimize the spectral response. The results demonstrate that the geometry of the tapered fiber can be regarded as an extra degree of freedom to the design of AO tunable modulators. For the particular case of 70 μm fiber modulator, high modulation depth of 60%, low insertion loss of 1.3 dB, and 40 nm of modulation bandwidth are achieved. If we compare this result with conventional non-tapered modulators [4, 15], we can appreciate the improvement achieved with the present scheme.

In Section 2, we start with the numerical modeling of acoustically induced LPG formed in a tapered structure. Then, in Section 3, we describe the experimental results and compare with numerical simulations to obtain insight into the effect of including tapered optical fiber in the AO device. Finally, our conclusions are summarized in Section 4.

2. Numerical modeling of the acousto-optic interaction in optical fibers

From the mechanical point of view, optical fibers are homogeneous cylinders immersed in air that have the ability to guide acoustic waves. These acoustic waves, depending on the displacement of their particles, are classified as flexural, longitudinal, or torsional waves. For the case of the flexural waves, a vibration is generated in one of the transverse directions of the fiber, which produces an asymmetric perturbation of the refractive index. In this section, the numerical technique to simulate the spectral response of the acoustically induced LPG is presented. Following the theory developed by Birks et al. in Ref. [3], the transmission matrix to numerically simulate the intermodal coupling between the core and cladding forward-propagating modes is reported.

The perturbation of the refractive index, $\Delta n(x,y)$, is described in terms of the acoustic wave as follows:

$$\Delta n(x, y) = \Delta n_0 \sin(\Omega t - Kz), \tag{1}$$

$$\Delta n_0 = n_0(1 - \chi)\, K^2\, u_0\, y, \tag{2}$$

where x, y, and z are the Cartesian coordinates. z represents the direction of propagation along the fiber for both the light and the acoustic wave, K is the acoustic wave number, Ω is the acoustic wave angular frequency, and Δn_0 is the perturbation amplitude (defined in Eq. (2)) with n being the refractive index of the fiber glass, χ is the photoelastic coefficient ($\chi = 0.22$ for silica), and u_0 is the amplitude of vibration of the flexural acoustic wave, which is assumed to vibrate in the direction of the y-axis. The coupling coefficient (κ) between the fundamental mode and a cladding mode is evaluated with the expression

$$\kappa = \frac{\omega}{2c} \iint_A \Delta n_0 E_{01}^{*} E_{lm}\, dx\, dy \tag{3}$$

where ω is the optical angular frequency, c is the speed of light, and E_{lm} is the normalized amplitude of the electric field of a fiber mode, being $l = 0$ and $m = 1$ the fundamental mode where * denotes the complex conjugate.

The fact that the perturbation has an explicit dependence on the transverse coordinate y implies a sinusoidal dependence with the azimuthal coordinate ϕ, $y = r \sin\phi$, which leads to an antisymmetric distribution of the perturbation in the cross section A of the fiber. Because the field distribution of the fundamental mode (LP_{01}) is symmetrical, to prevent the coupling coefficient from being null, the energy coupling is restricted to the cladding modes with asymmetric distribution. The modes that feature this asymmetric distribution feature are the LP_{1m} modes. Thus, there will be an intermodal coupling between the mode LP_{01} and the asymmetric modes of the family LP_{1m}. The total optical field is a superposition of two modes, which correspond to the core and cladding forward-propagating modes with local amplitudes $A(z)$ and $B(z)$, respectively. These amplitudes satisfy the following pair of coupled-mode equations:

$$\frac{dA(z)}{dz} = -j\,\kappa\, B(z)\, e^{j\Omega t}\, e^{2j\delta z} \tag{4}$$

$$\frac{dB(z)}{dz} = -j\,\kappa\, A(z)\, e^{-j\Omega t}\, e^{-2j\delta z} \tag{5}$$

where $j = (-1)^{1/2}$; $\delta = (1/2)(\beta_{01}: \beta_{lm})$: (π/Λ) is the phase mismatch parameter; β_{01} and β_{lm} are the propagation constants for the fundamental core mode and a high-order cladding mode, respectively; and Λ is the wavelength of the acoustic wave ($K = 2\pi/\Lambda$). In the regime of low frequencies, Λ can be expressed as

$$\Lambda = \sqrt{\frac{\pi a c_{ext}}{f}}, \tag{6}$$

where a is the radius of the fiber, c_{ext} is the extensional velocity (5760 m/s for silica), and f is the acoustic frequency.

The mode amplitudes, and hence the electric fields, can be readily solved from Eqs. (4) and (5) in terms of a matrix F [16]:

$$\begin{bmatrix} A(z)e^{-j\beta_{01}z} \\ B(z)e^{-j\beta_{lm}z} \end{bmatrix} = F \begin{bmatrix} A(0) \\ B(0) \end{bmatrix}, \tag{7}$$

$$F = \begin{bmatrix} \left(\cos(\gamma z) - j\frac{\delta}{\gamma}\sin(\gamma z)\right)e^{-j(\overline{\beta}+\frac{\pi}{\Lambda})z} & \left(-\frac{\kappa}{\gamma}e^{jKz}\sin(\gamma z)\right)e^{-j(\overline{\beta}+\frac{\pi}{\Lambda})z} \\ \left(\frac{\kappa}{\gamma}e^{-jKz}\sin(\gamma z)\right)e^{-j(\overline{\beta}-\frac{\pi}{\Lambda})z} & \left(\cos(\gamma z) + j\frac{\delta}{\gamma}\sin(\gamma z)\right)e^{-j(\overline{\beta}-\frac{\pi}{\Lambda})z} \end{bmatrix}, \tag{8}$$

where $\gamma = \left(\delta^2 + \kappa^2\right)^{1/2}$ and $\overline{\beta} = 1/2\left(\beta_{01} + \beta_{lm}\right)$.

A tapered structure is modeled assuming an acoustically induced LPG composed of N cascaded of uniforms LPGs with different fiber diameters and lengths L_i, as depicted in **Figure 1**.

In each section, since both the velocity (v) and the amplitude of the flexural wave depend on the fiber radius, the acoustic wavelength (Λ) and propagation constant (β) will have to be computed for each section of fiber. The velocity and amplitude of the acoustic are expressed in function of the fiber radius as

$$\upsilon = \sqrt{\pi R c_{ext}\, f}, \tag{9}$$

$$u_0(R) = \frac{R_0}{R} u_0(R_0), \tag{10}$$

where R_0 is the original radius of the fiber previous to tapering and where we have assumed that the acoustic power remains constant along the fiber.

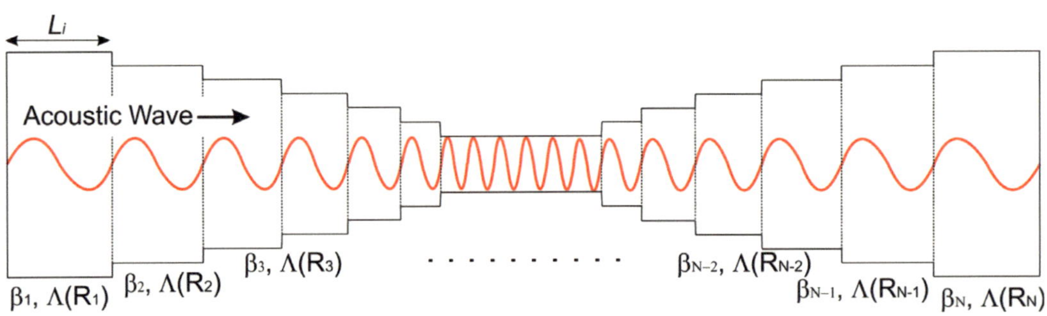

Figure 1. Simulation model for the tapered fiber structure.

The output from the *i*th section becomes the input to the (*i* + 1)th section of the fiber. The output from the *N*th section is therefore given by

$$\begin{bmatrix} A(L) \\ B(L) \end{bmatrix} = F_N \cdots F_2 F_1 \begin{bmatrix} A(0) \\ B(0) \end{bmatrix}, \tag{11}$$

where F_i is the *F*-matrix for the *i*th section, as defined by Eq. (8) with $z = L_i$, and $L = \Sigma L_i$ is the total length of the composite tapered fiber. The length L_i is chosen small enough to insure convergence of the numerical simulations.

Finally, the transmission response of the tapered AO device is modeled assuming that only the fundamental mode is launched into the fiber, i.e., $E_{01}(0) \neq 0$ and $E_{lm}(0) = 0$. Thus, the spectral dependence of the core mode is obtained computing $|E_{01}(L)|^2/|E_{01}(0)|^2$ for each optical wavelength. Previously, the specific cladding mode involved in the experiment will be identified after the theoretical analysis of the phase matching conditions of different cladding modes at the frequency of the acoustic wave.

3. Acousto-optic devices based on flexural waves

In this section we present experimental results on three different types of AO devices based on acoustic flexural waves. The first device is a conventional acousto-optic tunable attenuation filter, whose transmission characteristics are modeled with the previous transfer matrix analysis. Next, the spectral response of the attenuation filter is improved by the use of tapered optical fibers. The results demonstrate, as it will be shown in Section 3.2, that the geometry of the taper transitions can be regarded as an extra degree of freedom for the design of AO devices. A significant improvement in the spectral bandwidth could be achieved by the proper selection of the fiber transition. Finally, and with the objective of expanding the application of the AO tunable attenuation filters (AOTAF) to an amplitude modulator, acoustic reflection was intentionally induced. Thus, a standing acoustic wave is generated which produces an amplitude modulation at twice the acoustic frequency. Our approach permits the implementation of broad modulation bandwidth (40 nm), high modulation depth (60%), and low optical loss (1.3 dB) in a 70-μm configuration, operating in the MHz frequency range. These experimental results are within the best results reported in the framework of in-fiber AO modulators.

3.1. The acousto-optic tunable attenuation filter

A schematic view of the experimental setup to investigate the acousto-optic interaction in optical fibers is depicted in **Figure 2**. The AO device consists of a radio frequency (RF) source, a piezoelectric disk (PD), an aluminum concentrator horn, and an uncoated optical fiber. The PD is excited by the RF source with a sinusoidal waveform to produce an acoustic wave that is transmitted into the fiber through the aluminum horn. To prevent attenuation of the acoustic wave, the fiber is stripped of the outer polymer jacket; and to avoid unwanted acoustic reflections, the acoustic wave is damped at both ends of the uncoated fiber. The length (*L*) of the uncoated optical fiber was selected to be 24 cm. At this length, it is relatively easy to obtain

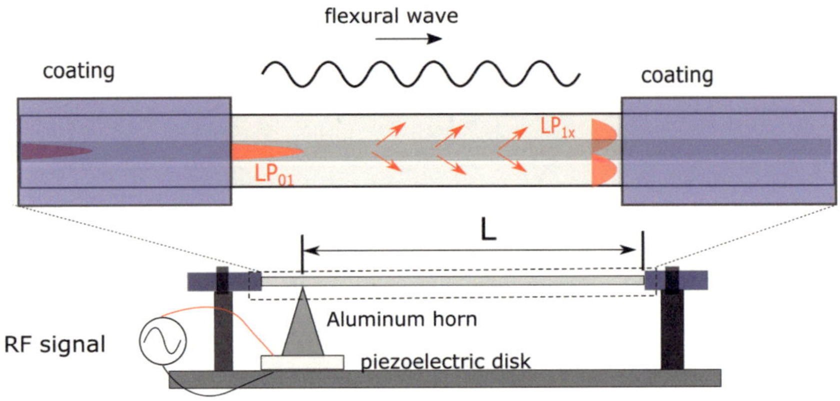

Figure 2. Acousto-optic attenuation filter.

a maximum transfer of energy between the core mode and some specific cladding mode of the fiber. Commercially available corning LEAF fiber was used in the experiments to implement AO devices.

Initial experiments were prepared to obtain the flexural wave frequencies that satisfy phase-matching condition between the LP_{01} mode and some specific higher-order cladding mode LP_{1m}. The objective is to identify the AO resonances in the experiments to ensure good agreement between numerical simulations and the observed AO interaction. The AO resonances and the spectral response were measured with a superluminescent diode light source (SLD) coupled to the optical fiber that provides an optical spectrum from 1450 to 1650 nm, and the transmission signal is monitored and recorded by an optical spectrum analyzer (OSA). Experimental results of the AO resonances are shown in **Figure 2** for a range of optical wavelengths between 1450 and 1650 nm. Working in the frequency range from 1.6 to 2.8 MHz, LP_{01}-LP_{11}, LP_{01}-LP_{12}, and LP_{01}-LP_{13} intermodal couplings were observed. The dotted line in **Figure 3(a)** indicates the acoustic frequency of 2.33 MHz, at which three AO resonances were observed with resonant optical wavelengths of 1473, 1537, and 1618 nm, respectively; see **Figure 3(b)**. The voltage applied to the piezoelectric disk (V_{PD}) was 33 V (whenever we refer to voltages, it is a peak-to-peak measurement). Beyond this voltage the response of the piezoelectric disk degrades due to an excessive heating. Most of our experiments were carried out using voltages around 30 V, where a stable and linear response of the PD was observed.

The strongest mode coupling was found at the acoustic frequency of 2.005 MHz, corresponding to the LP_{01}-LP_{12} intermodal coupling. The spectral dependence of the transmission is shown in **Figure 4**, where experimental (scatter points) and simulation (solid line) results are included. The AO resonance exhibits a maximum rejection efficiency of 15 dB at the optical resonant wavelength of 1596 nm and a 3-dB stopband bandwidth of 1.9 nm. The spectral dependence is calculated using Eq. (8), after computing the modal indexes of the fundamental and high-order cladding modes using a standard boundary-value method, applied to a step index fiber. Material dispersion was estimated assuming a binary silica optical fiber. Experimental results were found

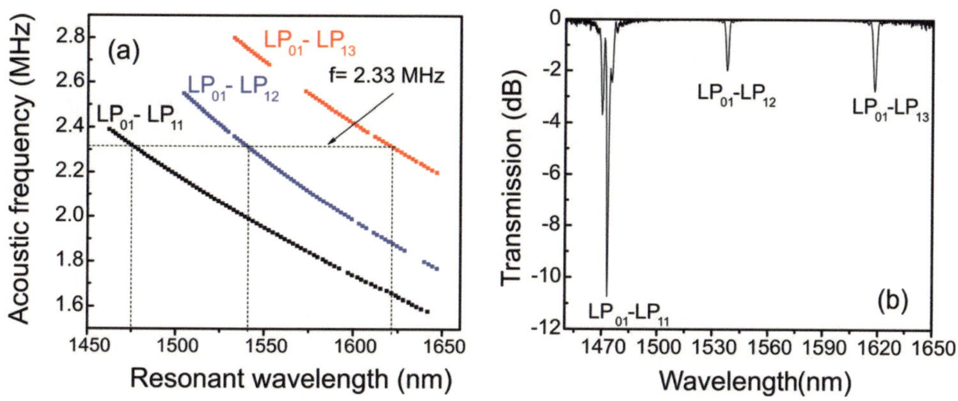

Figure 3. Acousto-optic resonances. (a) Acoustic frequency versus resonant optical wavelengths. (b) Transmission response of the AOTAF at the applied frequency of 2.33 MHZ.

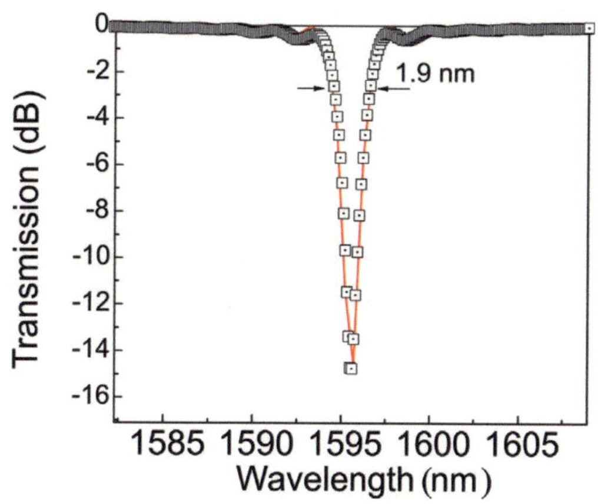

Figure 4. Transmittance spectrum at the acoustic frequency of 2.005 Mhz. Scatter points indicate the experimental results, and solid line indicates the simulated transmission.

to be in agreement with simulations by using Sellmeier coefficients for a silica glass doped with GeO_2 in the core and pure silica glass in the cladding [17, 18]. Experimental results were best fitted with a GeO_2 concentration of around 3.1%, maintaining constant the core radius at 9.3 μm. Simulation parameters were calculated to be closest to the LEAF fiber datasheet.

3.2. The acousto-optic tunable attenuation filter with a double-ended tapered fiber

In this section our purpose is to report an improved scheme of AO tunable attenuation filter based on the implementation of thin optical fibers. The proposed configuration combines the advantages of intermodal coupling in a double-ended tapered fiber. The fiber taper was prepared using a standard fusion and pulling technique, and it was stripped of the outer polymer jacket to prevent the attenuation of the acoustic wave.

In resonance, the resonant wavelength (λ_R) is directly proportional to the root square of the fiber radius R, expressed by the following relation:

$$\lambda_R = (n_{co} - n_{cl})\sqrt{\frac{\pi C_{ext} R}{f}}, \tag{12}$$

where n_{co} and n_{cl} represent the effective indexes of the fundamental core mode and the cladding mode, respectively. From Eq. (12) it is clear that a small variation in R produces a shift in the resonant wavelength λ_R. Hence, by imposing a gradual reduction in the fiber diameter, via the tapered fiber, a gradual shift of the resonant wavelength is expected, contributing to enrich the spectral response of the AO device. Furthermore, since the AO effect takes place into a thinner optical fiber, it produces a strong acousto-optic interaction that leads into a more efficient intermodal coupling [19, 20]. Experiments were realized with the same scheme to that shown in **Figure 2** but including a tapered fiber. The fiber was tapered down to a fiber diameter of 80 μm. The length of the tapered fiber is 23.7 cm long, and it was fabricated maintaining relatively long decaying exponential transition profiles. The tapered fiber was composed of 12.5-cm-long uniform taper waist and 5.6-cm-long exponential transitions. **Figure 5** shows the spectral response of the AO interaction. For comparison purposes, and to illustrate the improvement in the spectral response, the transmission spectra for a non-tapered optical fiber are included. In both cases, the acoustic frequency was selected to produce the maximum transfer of energy at a resonant wavelength around 1530 nm. Acoustic frequencies were fixed at 1.23 and 2.49 MHz for the tapered and non-tapered fiber, respectively.

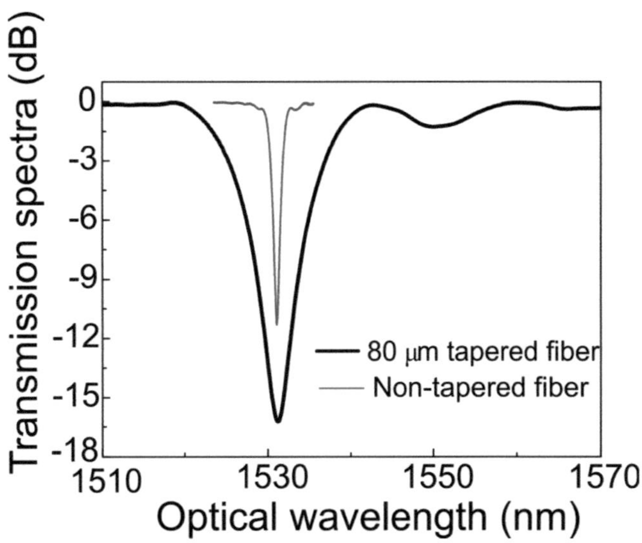

Figure 5. Transmission response of an 80-μm tapered fiber (black line). The blue line indicates the transmission response for a non-tapered optical fiber. The acoustic frequencies were fixed to 1.23 and 2.49 MHz, respectively.

The spectral response of the 80-μm tapered fiber exhibits a 3-dB optical bandwidth of 11.68 nm with a minimum transmittance of −16 dB at the resonant optical wavelength of 1531.2 nm. If we compare this result with the spectral response of the non-tapered fiber (12 dB of attenuation and 1.5 nm of optical bandwidth), the improvement achieved by implementing tapered optical fibers is clear.

In order to illustrate how the geometry of the tapered fiber can be used as an extra degree of freedom in the enhancement of the spectral response, we focus our attention on the effects of large tapered transitions as a mechanism to shape the transmission spectrum. The fibers were tapered down to fiber diameters of 80, 70, and 65 μm, respectively. All of them fabricated with long decaying exponential profiles and uniform waists of around ~10 cm in length. The dimensions of the tapered fibers were selected to produce a spectral response that is dependent on the taper transition. **Table 1** shows the parameters of the tapered fibers used in the experiments.

The transmission response of the AO interaction is shown in **Figure 6** for the three tapered fibers. The acoustic frequency was selected to produce a maximum attenuation resonance around 1550 nm, and that corresponded to 0.540, 1.194, and 1.207 MHz for the three tapers of 80, 70, and 65 μm, respectively. The rejection efficiency produced by these tapers was measured as 18, 6.2, and 4 dB, respectively. The red solid line in **Figure 6** indicates the simulated spectral response.

In these experiments, the notch bandwidth undergoes an improved spectral response that is dependent on the fiber diameter. For example, the bandwidth increases from 9 to 45 nm when the waist diameter is reduced from 80 to 70 μm, respectively, and then, the bandwidth reduces to 34 nm for the device with 65-μm waist diameter. This variation in the optical bandwidth can be associated to the spectral contributions of the taper waist and the taper transitions, as it will be

Waist diameter (μm)	Transition length (cm)	Waist length (cm)
80	5.06	11
70	5.90	10
65	5.98	9

Table 1. Parameters of the tapered fibers used in the experiments.

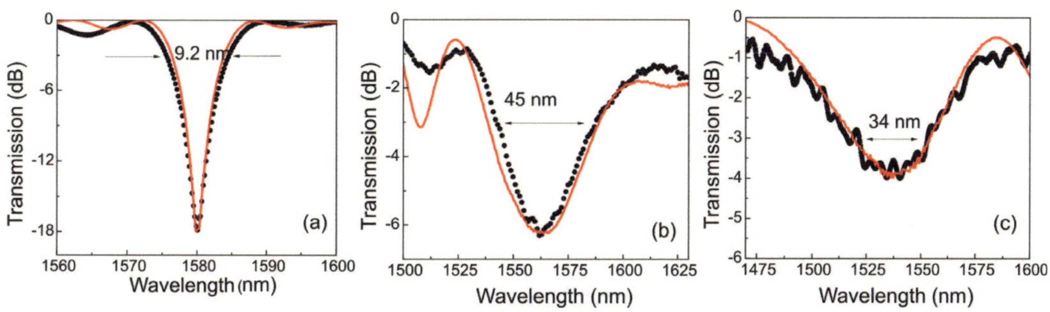

Figure 6. Transmission spectra for three different tapered fibers at different acoustic frequencies: (a) 80 μm, (b) 70 μm, and (c) 65 μm waist diameters. In all cases: simulations (solid curve) and experimental values (scatter points). Arrows indicate the spectral bandwidth at −3 dB.

shown below. By comparing these results with a non-tapered single-mode fiber, 1.5 nm of 3-dB optical bandwidth (see **Figure 5**), we observe a 6× to a 30× improvement in the optical bandwidth.

From these results (**Figure 6**) we conclude that our numerical simulations are realistic. Therefore, numeric simulation can be used to gain insight into the AO effect along the tapered optical fiber. For this purpose we have computed the contributions of isolated taper transitions and taper waist. Numerical simulations for the waist and the two taper transitions are shown in **Figure 7** for the three tapered fibers used in the experiments. For the device with 80-μm diameter in **Figure 7(a)**, the greatest contribution comes from the waist region of the device. Then, for the device with 70-μm waist diameter in **Figure 7(b)**, we can observe that the broadening of the notch is enhanced simultaneously by the contribution of both the taper waist and the taper transitions. Finally, for the device with 65-μm waist diameter, **Figure 7(c)**, the role of the taper transitions is the dominant to produce the spectral broadening.

Numerical results demonstrate that tapered transitions can play an important role to improve the optical bandwidth of AO devices. In order to obtain the greatest contribution, the response of taper transitions must be similar to the resonant wavelength of the taper waist. In this way, both elements contribute simultaneously to shape the spectral response of the device. We should also comment that the final transmission of the device is not the simple addition of isolated transmission sections, since a proper concatenation of the taper sections takes into account the amount of power already coupled in the previous sections and the phase accumulated in the each mode.

3.3. The acousto-optic amplitude modulator

An important characteristic of the acousto-optic effect occurs when acoustic reflection is induced. Under the effect of a standing flexural wave, the AOTAF can be operated as an amplitude modulator. With the objective of expanding the application of the filter to a broad bandwidth acousto-optic modulator (AOM), we take advantage of the improvements achieved in the spectral response by implementing tapered optical fibers. By firmly clamping one end of the fiber, opposite to the aluminum horn, a standing flexural acoustic wave is generated, and the transmission of the filter can be converted into an amplitude modulation. **Figure 8** illustrates the conversion of the filter into a modulator by the effect of

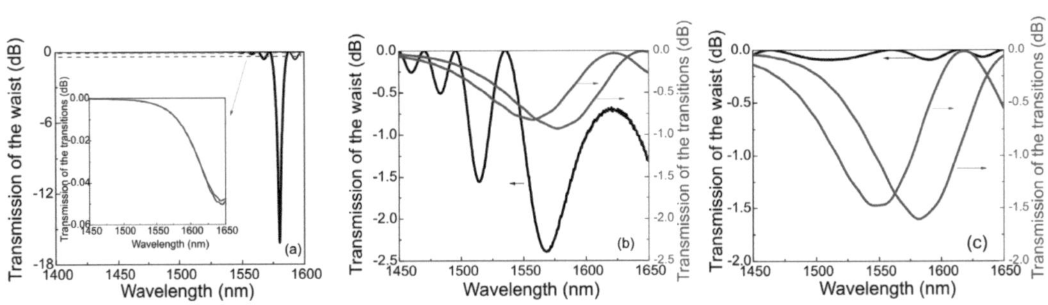

Figure 7. Numerical simulations of the spectral response considering the taper transitions and the taper waist as independent parts in the tapered fibers of (a) 80 μm, (b) 70 μm, and (c) 65 μm, respectively.

the standing wave. In this case, the device exhibits a transmission which oscillates in time as a result of the instantaneous perturbation generated by the standing flexural wave. Hence, the transmission is amplitude modulated at a frequency two times the frequency of the acoustic wave.

The AO modulator consists of a 24-cm-long tapered optical fiber, which is composed of two transition sections of 6.38 cm and a uniform waist of 70 μm with 11 cm in length. **Figure 9(a)** shows the spectral dependence including the maximum and minimum transmission for a RF signal applied to the piezoelectric disk of 1.1565 MHz and 10.2 V. From this result we observe a maximum attenuation of 9 dB at the resonant optical wavelength of 1541 nm. At this wavelength the intermodal coupling produces the maximum transfer of energy between the core and cladding modes and consequently the maximum amplitude modulation. The measurements in **Figure 9** were performed with a tunable laser diode (1520–1570 nm) by tuning the wavelength around the resonant wavelength and detecting the transmitted light in a standard oscilloscope. **Figure 9(b)** shows the transmitted light as a function of time at the resonant optical wavelength, for the same conditions described in **Figure 9(a)**. This result demonstrate an amplitude modulation at 2.313 MHz, which is two times the acoustic frequency used in the experiments (1.1565 MHz). The fact that the reflection coefficient for the acoustic wave is not 1 makes the maximum transmission to be slightly below the reference level, i.e., the

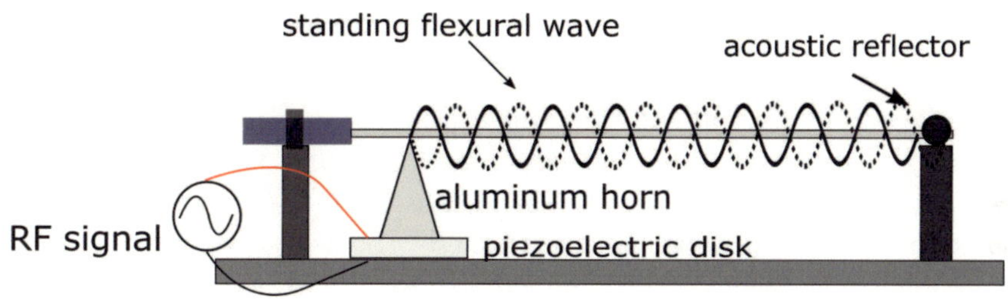

Figure 8. Acousto-optic amplitude modulator.

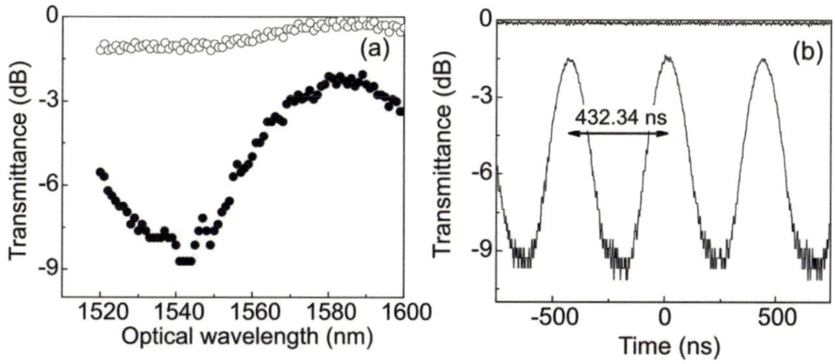

Figure 9. Transmission response of the AO modulator. (a) Maximum (solid points) and minimum (open points) transmission as a function of wavelength around the resonance located at 1541 nm. (b) Oscilloscope trace of transmitted light at the resonant optical wavelength; the reference level is the 0 dB line. In both cases f = 1.1565 MHz and V_{PD} = 10.2 V.

transmission of the fiber when no acoustic wave propagates. Therefore, the maximum transmission determines the insertion loss of the AOM at a given wavelength, RF frequency, and voltage. The difference between the maximum transmission and the minimum determines the modulation depth. From the results presented in **Figure 9**, we emphasize a strong modulation depth (60.5%), together with a low insertion loss (1.3 dB) and a broad operation bandwidth (~ 40 nm). By comparing the optical bandwidth with a non-tapered AOM [15] (1.5 nm bandwidth), a 26.67× improvement is achieved just by a small amount of reduction in the fiber diameter.

For practical applications, the modulator has a number of specific characteristics that require to be properly analyzed. First, we measured the modulation depth as a function of the optical wavelength, when both the acoustic frequency and the RF voltage were fixed. **Figure 10(a)** shows the modulation dependence around the resonant wavelength of 1541 nm when f = 1.1565 MHz and V_{PD} = 10.2 V are fixed. At the resonant wavelength, the modulation depth is maximal, and symmetrically decreases for longer and shorter wavelengths. The measured 3-dB bandwidth of the AO modulator is estimated as 40 nm, with a maximum modulation depth of 60.5%. On the other hand, since an acoustic resonator is also formed, **Figure 10(b)** shows the modulation depth versus the detuning frequency Δf when λ_R and V_{PD} are maintained at 1545 nm and V_{PD} = 10. 2 V, respectively. The center frequency in **Figure 10(b)** corresponds to 1.1565 MHz. At this frequency the modulation depth is maximal, and it drops gradually to values close to zero around frequencies of ±2 kHz. For longer and shorter frequencies, the transmission oscillates and decays to values near to zero for frequencies around ±6 kHz. Therefore, the proper operation of the AOM is determined by the acoustic frequency f, which is selected to achieve the maximum modulation depth.

From a practical point of view, the AOM may find practical applications as active mode locker in all-fiber laser for ultrashort pulse generation. As the results demonstrate, the modulation bandwidth could be as broad as the Erbium band emission (~40 nm), and it could be operated with modulation depths higher that 50%. Beside these benefits, the low insertion losses (< 2 dB) and the inherent advantages of being an all-fiber device should be mentioned. Further improvements in efficiency and interaction length could be possible by properly selecting the geometry of the tapered optical fiber in the modulator.

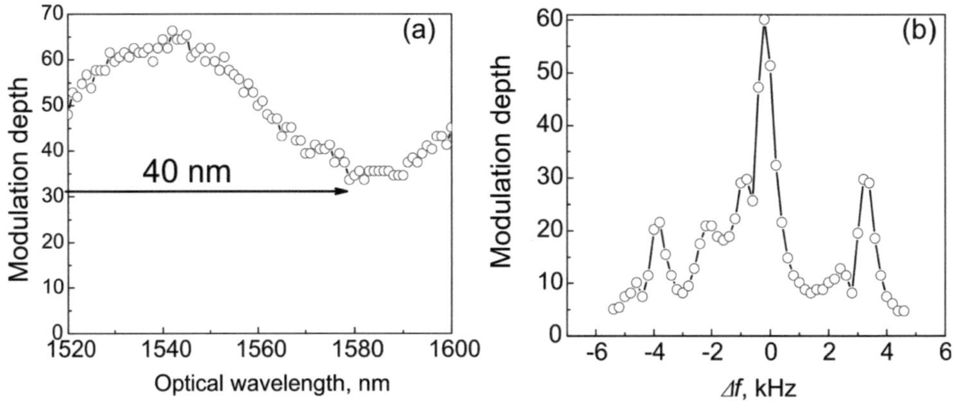

Figure 10. Characteristics of the AO modulation. (a) Modulation depth around the resonant wavelength of 1541 nm when f = 1.1565 MHz and V_{PD} = 10.2 V are fixed. (b) Modulation depth versus the detuning frequency Δf when λ_R and V_{PD} are maintained at 1541 nm and V_{PD} = 10.2 V, respectively. The central acoustic frequency in **(b)** is 1.1565 MHz.

4. Conclusions

This chapter demonstrated the potential of in-fiber acousto-optic interaction based on flexural acoustic waves for the design of all-fiber AO devices such as tunable attenuation filters and modulators. Based on the basic theory of AO interaction, the transmission matrix for simulating the acousto-optic response is developed. Numerical results demonstrate good agreement with experimental results. Thus, we can conclude that our numerical simulations are realistic, so we can use the simulation tool to gain insight into the AO interaction along the device. From experimental results, it is demonstrated that the geometry of the taper transitions can be regarded as an extra degree of freedom to the design of AO devices. Optical bandwidths of up to 45 nm are reported in a tapered fiber with a gradual reduction of the fiber. Additionally, under the effect of a standing flexural wave, conventional AO tunable attenuation filters can be operated as an amplitude modulator. As a particular case, an in-fiber AO modulator composed of a double-ended tapered fiber was reported. The fiber taper was prepared using a standard fusion and pulling technique, and it was tapered down to a fiber diameter of 70 μm. The device exhibits an amplitude modulation at 2.313 MHz, which is two times the acoustic frequency used (1.1565 MHz), a maximum modulation depth of 60.5%, 1.3 dB of insertion loss, and 40 nm of modulation bandwidth. From the point of view of implementation, the AOM is well suited for active mode locking in the ultrashort pulse regime.

Acknowledgements

This investigation has been financially supported by CONACyT grant 222476.

Author details

Miguel Ángel Bello Jiménez[1*], Gustavo Ramírez-Meléndez[1], Erika Hernández-Escobar[1], Andrés Camarillo-Avilés[1], Rosa López-Estopier[1,3], Olivier Pottiez[2], Cristian Cuadrado-Laborde[4], Antonio Díez[5], José L. Cruz[5] and Miguel V. Andrés[5]

*Address all correspondence to: miguel_bello@hotmail.com

1 Research Institute for Optics Communications (IICO), University of San Luis Potosí, San Luis Potosí, S.L.P., México

2 Optics Research Centre (CIO), León, GTO, Mexico

3 National Council on Science and Technology (CONACYT), Mexico, D. F., Mexico

4 Optical Metrology Lab, Institute of Physics Rosario (CONICET-UNR), Rosario, Argentina

5 Department of Applied Physics and Electromagnetism (ICMUV), University of Valencia, Burjassot, Valencia, Spain

References

[1] Kim BY, Blake JN, Engan HE, Shaw HJ. All-fiber acousto-optic frequency shifter. Optics Letters. 1986;**11**(6):389-391

[2] Östling D, Engan HE. Narrow-band acousto-optic tunable filtering in a two-mode fiber. Optics Letters. 1995;**20**(11):1247-1249

[3] Birks TA, Russell PSJ, Culverhouse DO. The acousto-optic effect in single-mode fiber tapers and couplers. Journal of Lightwave Technology. 1996;**14**(11):2519-2529

[4] Bello-Jiménez M, Cuadrado-Laborde C, Sáz-Rodríguez D, Díez A, Cruz JL, Andrés MV. Actively mode-locked fiber ring laser by intermodal acousto-optic modulation. Optics Letters. 2010;**35**(22):3781-3783

[5] Cuadrado-Laborde C, Bello-Jiménez M, Díez A, Cruz JL, Andrés MV. Long-cavity all-fiber ring laser actively mode locked with bandpass acousto-optic modulator. Optics Letters. 2014;**39**(1):68-71

[6] Alcusa-Sáes EP, Díez A, González-Herráez M, Andrés MV. Time-resolved acousto-optic interaction in single-mode optical fibers: Characterization of axial nonuniformities at the nanometer scale. Optics Letters. 2014;**39**(6):1437-1440

[7] Culverhouse DO, Richardson DJ, Birks TA, Russell PSJ. All-fiber sliding-frequency Er/Yb soliton laser. Optics Letters. 1995;**20**(23):2381-2383

[8] Liu WF, Russell PSJ, Dong L. Acousto-optic superlattice modulator using a fiber Bragg gratting. Optics Letters. 1997;**22**(19):1515-1517

[9] Jeon MY, Lee HK, Kim KH, Lee EH, WY O, Kim BY, et al. Harmonically mode-locked fiber laser with acousto-optic modulator in a Sagnac loop and Faraday rotating mirror cavity. Optics Communications. 1998;**149**(4-6):312-316

[10] Liu WF, Liu IM, Chung LW, Huang DW, Yang CC. Acoustic-induced switching of the reflection wavelength in a fiber Bragg gratting. Optics Letters. 2000;**25**(18):1319-1321

[11] Zalvidea D, Russo NA, Duchowicz R, Delgado-Pinar M, Díez A, Cruz JL, et al. High-repetition rate acoustic-induced Q-switched all-fiber laser. Optics Communications. 2005;**244**:315-319

[12] Andrés MV, Cruz JL, Díez A, Pérez-Millán P, Delgado-Pinar M. Actively Q-Switched all-fiber lasers. Laser Physics Letters. 2008;**5**(2):93-99

[13] Cuadrado-Laborde C, Diez A, Delgado-Pinar M, Cruz JL, Andrés MV. Mode locking of an all-fiber laser by acousto-optic superlattice modulation. Optics Letters. 2009;**34**(7):1111-1113

[14] Cuadrado-Laborde C, Díez A, Andrés MV. Experimental study of an all-fiber actively mode-locked by standing-wave acousto-optic modulation. Applied Physics B. 2010;**99**(1-2):95-99

[15] Bello-Jiménez M, Cuadrado-Laborde C, Diez A, Cruz JL, Andrés MV. Experimental study of an actively mode-locked fiber ring laser based on in-fiber amplitude modulation. Applied Physics B. 2001;**105**(2):269-276

[16] Erdogan T. Fiber gratting spectra. Journal of Lightwave Technology. 1997;**15**(8):1277-1294

[17] Jeong H, Oh K. Theoretical analysis of cladding-mode wave-guide dispersion and its effects on the spectra of long-period fiber grattings. Journal of Lightwave Technology. 2003;**21**(8):1838-1845

[18] Flemming JW. Material dispersion in lightguide glasses. Electronic Letters. 1978;**14**(11): 326-328

[19] Li Q, Liu X, Peng J, Zhou B, Lyons ER, Lee HP. Highly efficient acoustooptic tunable filter based on cladding erched single-mode fiber. IEEE Photonics Technology Letters. 2002;**14**(3): 337-339

[20] Abrishamian F, Nagai S, Sato S, Imai M. Design theory and experiment of acousto-optical tunable filter by use of flexural waves applied to thin optical fiber. Optical and Quantum Electronics. 2008;**40**(9):665-676

Chapter 6

Wave Propagation in Porous Materials

Zine El Abiddine Fellah, Mohamed Fellah,
Claude Depollier, Erick Ogam and Farid G. Mitri

Additional information is available at the end of the chapter

http://dx.doi.org/10.5772/intechopen.72215

Abstract

This chapter provides different models for the acoustic wave propagation in porous materials having a rigid and an elastic frames. The direct problem of reflection and transmission of acoustic waves by a slab of porous material is studied. The inverse problem is solved using experimental reflected and transmitted signals. Both high- and low-frequency domains are studied. Different acoustic methods are proposed for measuring physical parameters describing the acoustic propagation as porosity, tortuosity, viscous and thermal characteristic length, and flow resistivity. Some advantages and perspectives of this method are discussed.

Keywords: acoustic porous materials, porosity, tortuosity, viscous and thermal charactertistic lengths, fractional derivatives

1. Introduction

More than 50 years ago, Biot [1, 2] proposed a semi-phenomenological theory which provides a rigorous description of the propagation of acoustic waves in porous media saturated by a compressible viscous fluid. Due to its very general and rather fundamental character, it has been applied in various fields of acoustics such as geophysics, underwater acoustics, seismology, ultrasonic characterization of bones, etc. Biot's theory describes the motion of the solid and the fluid, as well as the coupling between the two phases. The loss of acoustic energy is due mainly to the viscosity of the fluid and the relative fluid-structure movement. The model predicts that the acoustic attenuation, as well as the speed of sound, depends on the frequency and elastic constants of the porous material, as well as porosity, tortuosity, permeability, etc. The theory predicts two compressional waves: a fast wave, where the fluid and solid move in phase, and a slow wave where fluid and solid move out of phase. Johnson et al. [3] introduced the concept of tortuosity or dynamic permeability which has better described the viscous losses between fluid and structure in both high and low frequencies.

Air-saturated porous materials such as plastic foams or fibrous materials are widely used in passive control and noise reduction. These materials have interesting acoustic properties for sound absorption, and their use is quite common in the building trade and automotive and aeronautical fields. The determination of the physical parameters of the medium from reflected and transmitted experimental data is a classical inverse scattering problem.

Pulse propagation in porous media is usually modeled by synthesizing the signal via a Fourier transform of the continuous wave results. On the other hand, experimental measurements are usually carried out using pulses of finite bandwidth. Therefore, direct modeling in the time domain is highly desirable [4–10]. The temporal and frequency approaches are complementary for studying the propagation of acoustic signals. For transient signals, the temporal approach is the most

Figure 1. Air-saturated plastic foam.

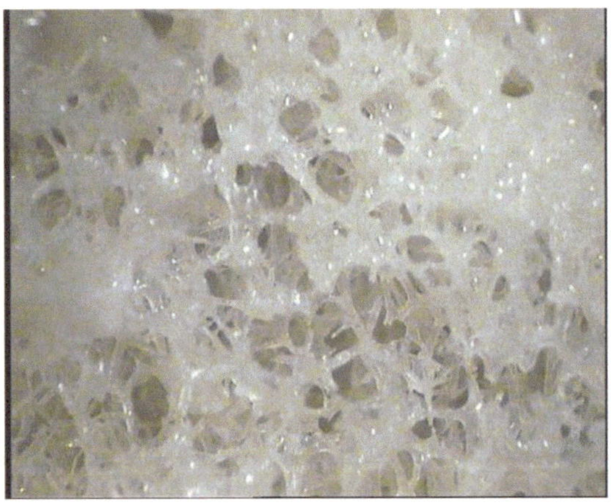

Figure 2. Human cancellous bone sample.

appropriate because it is closer to the experimental reality and the finite duration of the signal. However, for monochromatic harmonic signals, the frequency approach is the most suitable [11].

Fractional calculus has been used in the past by many authors as an empirical method to describe the viscoelastic properties of materials (e.g., see Caputo [12] and Bagley and Torvik [13]). The fact that acoustic attenuation, stiffness, and damping in porous materials are proportional to the fractional powers of frequency [4, 5, 7, 9, 10] suggests that fractional-order time derivatives could describe the propagation of acoustic waves in these materials.

In this chapter, acoustic wave propagation in porous media is studied in the high- and the low-frequency range. The direct and inverse scattering problems are solved for the mechanical characterization of the medium. The general Biot model applied to porous materials having elastic structure is treated, and also the equivalent fluid model, used for air-saturated porous materials (**Figures 1** and **2**).

2. Porous materials with elastic frame

In porous media, the equations of motion of the frame and fluid are given by the Euler equations applied to the Lagrangian density. Here, **u** and **U** are the displacements of the solid and fluid phases. The equations of motion are given by [1, 2]

$$\rho_{11}\frac{\partial^2 \mathbf{u}}{\partial t^2} + \rho_{12}\frac{\partial^2 \mathbf{U}}{\partial t^2} = P\nabla.(\nabla \mathbf{u}.) + Q\nabla(\nabla.\mathbf{U}) - N\nabla \wedge (\nabla \wedge \mathbf{u}), \quad (1)$$

$$\rho_{12}\frac{\partial^2 \mathbf{u}}{\partial t^2} + \rho_{22}\frac{\partial^2 \mathbf{U}}{\partial t^2} = Q\nabla(\nabla.\mathbf{u}) + R\nabla(\nabla.\mathbf{U}), \quad (2)$$

where P, Q, and R are the generalized elastic constants, ϕ is the porosity, K_f is the bulk modulus of the pore fluid, K_s is the bulk modulus of the elastic solid, and K_b is the bulk modulus of the porous skeletal frame. N is the shear modulus of the composite as well as that of the skeletal frame. The equations which explicitly relate P, Q, and R to ϕ, K_f, K_s, K_b, and N are given by

$$P = \frac{(1-\phi)\left(1-\phi-\frac{K_b}{K_s}\right)K_s + \phi\frac{K_s}{K_f}K_b}{1-\phi-\frac{K_b}{K_s}+\phi\frac{K_s}{K_f}} + \frac{4}{3}N, \quad Q = \frac{\left(1-\phi-\frac{K_b}{K_s}\right)\phi K_s}{1-\phi-\frac{K_b}{K_s}+\phi\frac{K_s}{K_f}}, \quad R = \frac{\phi^2 K_s}{1-\phi-\frac{K_b}{K_s}+\phi\frac{K_s}{K_f}}.$$

ρ_{mn} is the "mass coefficients" which are related to the densities of solid (ρ_s) and fluid (ρ_f) phases by

$$\rho_{11} + \rho_{12} = (1-\phi)\rho_s, \qquad \rho_{12} + \rho_{22} = \phi\rho_f. \quad (3)$$

The Young modulus and the Poisson ratio of the solid E_s and ν_s and of the skeletal frame E_b and ν_b depend on the generalized elastic constant P, Q, and R via the relations:

$$K_s = \frac{E_s}{3(1-2\nu_s)}, \qquad K_b = \frac{E_b}{3(1-2\nu_b)}, \qquad N = \frac{E_b}{2(1+\nu_b)}. \tag{4}$$

The mass coupling parameter ρ_{12} between the fluid and solid phases is always negative:

$$\rho_{12} = -\phi\rho_f(\alpha_\infty - 1), \tag{5}$$

where α_∞ is the tortuosity of the medium. The damping of the acoustic wave in porous material is essentially due to the viscous exchanges between the fluid and the structure. To express the viscous losses, the dynamic tortuosity is introduced [3] $\alpha(\omega)$ given by

$$\alpha(\omega) = \alpha_\infty\left(1 - \frac{1}{jx}\sqrt{1 - \frac{M}{2}jx}\right) \quad \text{where} \quad x = \frac{\omega\alpha_\infty\rho_f}{\sigma\phi} \quad \text{and} \quad M = \frac{8k_0\alpha_\infty}{\phi\Lambda^2}. \tag{6}$$

where $j^2 = -1$, ω is the angular frequency, σ is the fluid resistivity, k_0 is the viscous permeability, and Λ is the viscous characteristic length given by Johnson et al. [3]. The ratio of the sizes of the pores to the viscous skin depth thickness $\delta = (2\eta/\omega\rho_0)^{1/2}$ gives an estimation of the parts of the fluid affected by the viscous exchanges. In this domain of the fluid, the velocity distribution is perturbed by the frictional forces at the interface between the viscous fluid and the motionless structure. At high frequencies, the viscous skin thickness is very thin near the radius of the pore r. The viscous exchanges are concentrated in a small volume near the surface of the frame $\delta/r \ll 1$. The expression of the dynamic tortuosity $\alpha(\omega)$ is given by [3]

$$\alpha(\omega) = \alpha_\infty\left(1 + \frac{2}{\Lambda}\left(\frac{\eta}{j\omega\rho_f}\right)^{1/2}\right), \tag{7}$$

The range of frequencies such that viscous skin thickness $\delta = (2\eta/\omega\rho_0)^{1/2}$ is much larger than the radius of the pores r

$$\frac{\delta}{r} \gg 1 \tag{8}$$

is called the low-frequency range. For these frequencies, the viscous forces are important everywhere in the fluid. When $\omega \to 0$, the expression of the dynamic tortuosity becomes

$$\alpha(\omega) \approx \alpha_0\left(1 + \frac{\eta\phi}{j\omega\alpha_0\rho_f k_0}\right), \tag{9}$$

α_0 is the low-frequency approximation of the tortuosity introduced by Lafarge in [14] and Norris [15]:

$$\alpha_0 = \frac{< \mathbf{v}(\mathbf{r})^2 >}{< \mathbf{v}(\mathbf{r})>^2} \tag{10}$$

where $\mathbf{v}(\mathbf{r})$ is the microscopic velocity. The angle brackets represent the average of the random variable over the sample of material. In the time domain, and in the high-frequency domain, the dynamic tortuosity (Eq. 7) $\alpha(\omega)$ acts as the operator, and its expression is given by [8]

$$\tilde{\alpha}(t) = \alpha_\infty \left(\delta(t) + \frac{2}{\Lambda} \left(\frac{\eta}{\pi \rho_f} \right)^{1/2} t^{-1/2} \right), \tag{11}$$

$\delta(t)$ is the Dirac function. In this model the time convolution of $t^{-1/2}$ with a function is interpreted as a semi-derivative operator according to the definition of the fractional derivative of order ν given by Samko et al. [16]:

$$D^\nu[x(t)] = \frac{1}{\Gamma(-\nu)} \int_0^t (t-u)^{-\nu-1} x(u) du, \tag{12}$$

where $0 \leq \nu < 1$ and $\Gamma(x)$ is the gamma function. A fractional derivative acts as a convolution integral operator and no longer represents the local variations of the function. The properties of fractional derivatives and fractional calculus are given by Samko et al. [16].

The introduction of the tortuosity operator $\tilde{\alpha}(t)$ (Eq. 11) in Biot's Eqs. (1) and (2) to describe the inertial and viscous interactions between fluid and structure will express the propagation equations in the time domain. When $\tilde{\alpha}(t)$ is used instead of α_∞ in Eqs. (1) and (2), the equations of motion (1) and (2) will be written as [17]

$$\int_0^t \tilde{\rho}_{11}(t-t') \frac{\partial^2 \mathbf{u}(t')}{\partial t^2} + \int_0^t \tilde{\rho}_{12}(t-t') \frac{\partial^2 \mathbf{u}(t')}{\partial t'^2} dt = P.\nabla(\nabla.\mathbf{u}(\mathbf{t})) + Q\nabla(\nabla.\mathbf{u}(t)) - N\nabla \wedge (\nabla \wedge \mathbf{u}(t)),$$

$$\int_0^t \tilde{\rho}_{12}(t-t') \frac{\partial^2 \mathbf{u}(t)(t')}{\partial t^2} + \int_0^t \tilde{\rho}_{22}(t-t') \frac{\partial^2 \mathbf{U}(t')}{\partial t^2} dt = Q\nabla(\nabla.\mathbf{u}(t)) + R\nabla(\nabla.\mathbf{U}(t)). \tag{13}$$

In these equations, the temporal operators $\tilde{\rho}_{11}(t)$, $\tilde{\rho}_{12}(t)$, and $\tilde{\rho}_{22}(t)$ represent the mass coupling operators between the fluid and solid phases and are given by

$$\tilde{\rho}_{11}(t) = (1-\phi)\rho_s + \phi\rho_f\left(\tilde{\alpha}(t) - 1\right), \qquad \tilde{\rho}_{12}(t) = -\phi\rho_f\left(\tilde{\alpha}(t) - 1\right), \qquad \tilde{\rho}_{22}(t) = \phi\rho_f\tilde{\alpha}(t),$$

where $\tilde{\alpha}(t)$ is given by Eq. (11).

The wave equations of dilatational and rotational waves can be obtained using scalar and vector displacement potentials, respectively. Two scalar potentials for the frame and the fluid, Φ_s and Φ_f, are defined for compressional waves giving

$$\begin{pmatrix} \rho_{11} \frac{\partial^2}{\partial t^2} + A \frac{\partial^{3/2}}{\partial t^{3/2}} - P\Delta & \rho_{12} \frac{\partial^2}{\partial t^2} - A \frac{\partial^{3/2}}{\partial t^{3/2}} - Q\Delta \\ \rho_{12} \frac{\partial^2}{\partial t^2} - A \frac{\partial^{3/2}}{\partial t^{3/2}} - Q\Delta & \rho_{22} \frac{\partial^2}{\partial t^2} + A \frac{\partial^{3/2}}{\partial t^{3/2}} - R\Delta \end{pmatrix} \begin{pmatrix} \tilde{\Phi}_s(t) \\ \tilde{\Phi}_f(t) \end{pmatrix} = 0. \tag{14}$$

where $A = \frac{2\phi\rho_f\alpha_\infty}{\Lambda}\sqrt{\frac{\eta}{\rho_f}}$, Δ is the Laplacian, and $\frac{\partial^{3/2}}{\partial t^{3/2}}$ represents the fractional derivative following the definition given by Eq. (12).

Two distinct longitudinal modes called fast and slow waves are obtained by the resolution of the eigenvalue problem of the matrix of Biot (Eq. (14)). On a basis of fast and slow waves $\Phi_1(t)$ and $\Phi_2(t)$, one can have

$$\Delta\begin{pmatrix}\Phi_1(t)\\ \Phi_2(t)\end{pmatrix} = \begin{pmatrix}\tilde{\lambda}_1(t) & 0\\ 0 & \tilde{\lambda}_2(t)\end{pmatrix}\begin{pmatrix}\Phi_1(t)\\ \Phi_2(t)\end{pmatrix}, \tag{15}$$

where $\tilde{\lambda}_1(t)$ and $\tilde{\lambda}_2(t)$ are the "eigenvalue operators" of the Biot matrix (Eq. (14)). Their expressions are given by

$$\tilde{\lambda}_i(t) = C_i\frac{\partial^2}{\partial t^2} + D_i\frac{\partial^{3/2}}{\partial t^{3/2}} + G_i\frac{\partial}{\partial t}, \qquad i = 1, 2, \tag{16}$$

Their corresponding eigenvectors are

$$\tilde{\mathfrak{J}}_i(t) = A_i + \frac{B_i}{\sqrt{\pi t}}, \qquad i = 1, 2, \tag{17}$$

where

$$C_i = \frac{1}{2}\left(\tau_1 + (-1)^i\sqrt{\tau_1^2 - 4\tau_3}\right), \quad D_i = \frac{1}{2}\left(\tau_2 + (-1)^i\frac{\tau_1\tau_2 - 2\tau_4}{\sqrt{\tau_1^2 - 4\tau_3}}\right),$$

$$G_i = (-1)^i.\frac{1}{4}\left(\frac{\tau_2^2}{\sqrt{\tau_1^2 - 4\tau_3}} - \frac{(\tau_1\tau_2 - 2\tau_4)^2}{2(\tau_1^2 - 4\tau_3)^{3/2}}\right), A_i = \frac{\tau_1 - 2\tau_5 + (-1)^i\sqrt{\tau_1^2 - 4\tau_3}}{2\tau_7},$$

$$B_i = \frac{1}{4\tau_7^2}\left[\left(\tau_2 - 2\tau_6 + (-1)^i\frac{\tau_1\tau_2 - 2\tau_4}{\sqrt{\tau_1^2 - 4\tau_3}}\right)2\tau_7 + \left(\tau_1 - 2\tau_5 - \sqrt{\tau_1^2 - 4\tau_3}\right).2\tau_6\right], \quad i = 1, 2,$$

and

$$\tau_1 = R'\rho_{11} + P'\rho_{22} - 2Q'\rho_{12}, \quad \tau_2 = A(P' + R' + 2Q'), \quad \tau_3 = \left(P'R' - Q'^2\right)\left(\rho_{11}\rho_{22} - \rho_{12}^2\right),$$

$$t\ \tau_4 = A\left(P'R' - Q'^2\right)(\rho_{11} + \rho_{22} - 2\rho_{12}), \quad \tau_5 = (R'\rho_{11} - Q'\rho_{12}), \quad \tau_6 = A(R' + Q'),$$

$$\tau_7 = (R'\rho_{12} - Q'\rho_{22}).$$

Coefficients R', P', and Q' are given by

$$R' = \frac{R}{PR - Q^2}, \qquad Q' = \frac{Q}{PR - Q^2}, \qquad \text{and} \quad P' = \frac{P}{PR - Q^2}.$$

The fast and slow waves Φ_1 and Φ_2 are obeying to the following propagation equations along the x axis:

$$\frac{\partial^2 \Phi_i(x,t)}{\partial x^2} - \frac{1}{v_i^2}\frac{\partial^2 \Phi_i(x,t)}{\partial t^2} - h_i \frac{\partial^{3/2} \Phi_i(x,t)}{\partial t^{3/2}} - d\frac{\partial \Phi_i(x,t)}{\partial t} = 0, \qquad i = 1,2, \tag{18}$$

where the coefficients v_i, h_i $(i = 1,2)$, and d are constants, respectively, given by

$$v_i = \frac{2}{\sqrt{\sqrt{\tau_1^2 - 4\tau_3} + (-)^i \tau_1}}, h_i = \frac{1}{2}\left(\tau_2 + (-1)^i \frac{\tau_1\tau_2 - 2\tau_4}{\sqrt{\tau_1^2 - 4\tau_3}}\right), \quad i = 1,2$$

and

$$d = -\frac{1}{4}\left(\frac{\tau_2^2}{\sqrt{\tau_1^2 - 4\tau_3}} - \frac{(\tau_1\tau_2 - 2\tau_4)^2}{2(\tau_1^2 - 4\tau_3)^{3/2}}\right),$$

where Eq. (18) is a fractional propagation equations [17] in time domain of the fast and slow waves, respectively. These equations describe the attenuation and the spreading of the temporal signal propagating inside the porous material. These fractional propagation equations have been solved and well-studied in the case of rigid porous materials using the equivalent fluid model.

3. Porous materials with rigid frame

In the acoustics of porous media, two situations can be distinguished: elastic and rigid frame materials. In the first case, the Biot [1, 2] theory is best suited. In the second case, the acoustic wave cannot vibrate the structure. The equivalent fluid model is then used, in which the acoustic wave propagates inside the saturating fluid [8, 11]. The equations for the acoustics in the equivalent fluid model are given by

$$\rho \frac{\partial^2 U_i}{\partial t^2} = -\nabla_i p, \qquad p = -K_f \nabla . \mathrm{U}. \tag{19}$$

In these relations, p is the acoustic pressure. The first equation is the Euler equation, and the second one is a constitutive equation obtained from the equation of mass conservation associated with the behavior (or adiabatic) equation. These equations can be obtained from the Biot Eqs. (1, 2) by canceling the solid displacement. Assuming that the porous medium studied is homogeneous and has a linear elasticity, we obtain easily the following wave equation (propagation along the x axis) for the acoustic pressure in a lossless porous material:

$$\frac{\partial^2 p(x,t)}{\partial x^2} - \left(\frac{\rho}{K_a}\right)\frac{\partial^2 p(x,t)}{\partial t^2} = 0. \tag{20}$$

In Eq. (20), the viscous and thermal losses that contribute to the sound damping in acoustic materials are not described. The thermal exchanges are generally negligible near viscous effects in the porous materials obeying to the Biot theory, this is not the case for air-saturated porous materials using the equivalent fluid model. To take into account the fluid-structure exchanges, the density and compressibility of the fluid are "renormalized" by the dynamic tortuosity $\alpha(\omega)$ and the dynamic compressibility $\beta(\omega)$, via the relations $\rho \to \rho\alpha(\omega)$ and $K_f \to K_f/\beta(\omega)$, giving the following wave equation in frequency domain (Helmholtz equation) for a lossy porous material:

$$\frac{\partial^2 p(x,t)}{\partial x^2} + \omega^2 \left(\frac{\rho\alpha(\omega)\beta(\omega)}{K_a} \right) p(x,t) = 0. \tag{21}$$

The thermal exchanges to the fluid compressions-dilatations are produced by the wave motion. The parts of the fluid affected by the thermal exchanges can be estimated by the ratio of a microscopic characteristic length of thermal skin depth thickness $\delta' = (2\eta/\omega\rho P_r)^{1/2}$ (η is the fluid viscosity; P_r is the Prandtl number).

The expression of the dynamic compressibility is given by

$$\beta(\omega) = \gamma - (\gamma - 1) / \left[1 - \frac{1}{jx'} \sqrt{1 - \frac{M'}{2} jx'} \right] \text{ where } x' = \frac{\omega \rho_f k'_0 P_r}{\eta \phi} \text{ and } M' = \frac{8k'_0}{\phi \Lambda'^2}. \tag{22}$$

where γ is the adiabatic constant, the magnitude k'_0 introduced by Lafarge [14] called thermal permeability by analogy to the viscous permeability, and Λ' is the thermal characteristic length. The low-frequency approximation of $\beta(\omega)$ [14] is given by

$$\beta(\omega) = \gamma + \frac{(\gamma - 1)\rho_f k_0{}' P_r}{\eta \phi j\omega}, \qquad \text{when} \qquad \omega \to 0. \tag{23}$$

where k'_0, which has the same size (area) that of Darcy's permeability of k_0, is a parameter analogous to the parameter k_0 but is adapted to the thermal problem.

In a high-frequency limit, Allard and Champoux [18] showed the following behavior of $\beta(\omega)$:

$$\beta(\omega) = 1 - \frac{2(\gamma - 1)}{\Lambda'} \left(\frac{\eta}{P_r \rho_f} \right)^{1/2} \left(\frac{1}{j\omega} \right)^{1/2}, \qquad \omega \to \infty. \tag{24}$$

Replacing $\alpha(\omega)$ and $\beta(\omega)$ given by Eqs (18) in Eq. (21), we obtain the following lossy equation for porous materials in the high-frequency domain:

$$\frac{\partial^2 p(x,t)}{\partial x^2} + \omega^2 \frac{\rho\alpha}{K_a} \left(1 - \sqrt{\frac{\eta}{\rho j\omega}} \left[\frac{2}{\Lambda} + \frac{2(\gamma - 1)}{\Lambda' \sqrt{\mathrm{Pr}}} \right] \right) p(x,t) + \left(\frac{D_1 - 1}{x} \right) \frac{\partial p(x,t)}{\partial x} = 0. \tag{25}$$

In the time domain (using the convention $\partial/\partial t \to -j\omega$), we obtain the following fractional propagation equation:

http://dx.doi.org/10.5772/intechopen.72215

$$\frac{\partial^2 p(x,t)}{\partial x^2} - \left(\frac{\rho\alpha}{K_a}\right)\frac{\partial^2 p(x,t)}{\partial t^2} - \frac{2\alpha\sqrt{\rho\eta}}{Ka}\left(\frac{2}{\Lambda} + \frac{2(\gamma-1)}{\Lambda'\sqrt{\Pr}}\right)\frac{\partial^{3/2} p(x,t)}{\partial t^{3/2}} = 0. \tag{26}$$

In this equation, the term $\frac{\partial^{3/2} p(x,t)}{\partial t^{3/2}}$ is interpreted as a semi-derivative operator following the definition of the fractional derivative of order ν, given by Samko and coll. [16]. The solution of the wave Eq. (26) with suitable initial and boundary conditions is by using the Laplace transform. F is the medium's Green function [9] given by

$$F(t,k) = \begin{cases} 0 & if \quad 0 \le t \le k \\ \Xi(t) + \Delta\int_0^{t-k} h(t,\xi)d\xi & \text{if} \quad t \ge k \end{cases} \tag{27}$$

with

$$\Xi(t) = \frac{b'}{4\sqrt{\pi}}\frac{k}{(t-k)^{3/2}}\exp\left(-\frac{b'^2k^2}{16(t-k)}\right), \tag{28}$$

where $h(\tau,\xi)$ has the following form:

$$h(\xi,\tau) = -\frac{1}{4\pi^{3/2}}\frac{1}{\sqrt{(\tau-\xi)^2-k^2}}\frac{1}{\xi^{3/2}}\int_{-1}^{1}\exp\left(-\frac{\chi(\mu,\tau,\xi)}{2}\right)(\chi(\mu,\tau,\xi)-1)\frac{\mu d\mu}{\sqrt{1-\mu^2}}, \tag{29}$$

$$\chi(\mu,\tau,\xi) = \left(\Delta\mu\sqrt{(\tau-\xi)^2-k^2} + b'(\tau-\xi)\right)^2/8\xi, b' = Bc_0^2\sqrt{\pi},$$

and $\Delta = b'^2$.

Let us consider a homogeneous porous material which occupies the region $0 \le x \le L$; the expressions of the reflection and transmission coefficients in the frequency domain are given by

$$R(\omega) = \frac{(1-D^2)\sinh(k(\omega)L)}{2Y(\omega)\coth(k(\omega)L) + (1+Y^2(\omega))\sinh(k(\omega)L)}, \tag{30}$$

$$T(\omega) = \frac{2Y(\omega)}{2Y(\omega)\coth(k(\omega)L) + (1+Y^2(\omega))\sinh(k(\omega)L)}, \tag{31}$$

where

$$Y(\omega) = \phi\sqrt{\frac{\beta(\omega)}{\alpha(\omega)}}, \quad \text{and} \quad k(\omega) = \omega\sqrt{\frac{\rho\alpha(\omega)\beta(\omega)}{K_a}},$$

These expressions are simplified by taking into account the reflections at the interfaces $x = 0$ and $x = L$; the expressions of the reflection and transmission operators are given in time domain by

$$\tilde{R}(t) = \frac{\sqrt{\alpha_\infty} - \phi}{\sqrt{\alpha_\infty} + \phi}\delta(t) - \frac{4\phi\sqrt{\alpha_\infty}\left(\sqrt{\alpha_\infty} - \phi\right)}{\left(\sqrt{\alpha_\infty} + \phi\right)^3} F\left(t, \frac{2L}{c}\right), \tag{32}$$

$$\tilde{T}(t) = \frac{4\phi\sqrt{\alpha_\infty}}{\left(\phi + \sqrt{\alpha_\infty}\right)^2} F\left(t + \frac{L}{c}, \frac{L}{c}\right). \tag{33}$$

where $\delta(t)$ is the Dirac function and F is the Green function of the medium given by Eq. (27). In the next sections, we will use the reflected and transmitted waves for solving the inverse problem in order to characterize the porous materials.

3.1. Ultrasonic measurement of porosity, tortuosity, and viscous and thermal characteristic lengths via transmitted waves

The experimental setup consists of two transducers broadband Ultran NCT202 with a central frequency of 190 kHz in air and a bandwidth of 6 dB extending from 150 to 230 kHz [19]. A pulser/receiver 5058PR Panametrics sends pulses of 400 V. The high-frequency noise is avoided by filtering the received signals above 1 MHz. Electronic interference is eliminated by 1000 acquisition averages. The experimental setup is shown in **Figure 3**. The inverse problem is to find the parameters α_∞, ϕ, Λ, and Λ' which minimize numerically the discrepancy function $U(\alpha_\infty, \phi, \Lambda, \Lambda') = \sum_{i=1}^{i=N}\left(p^t_{exp}(x, t_i) - p^t(x, t_i)\right)^2$, wherein $p^t_{exp}(x, t_i)_{i=1,2,\ldots n}$ is the discrete set of values of the experimental transmitted signal and $p^t(x, t_i)_{i=1,2,\ldots n}$ is the discrete set of values of the simulated transmitted signal predicted from Eq. (33). The least squares method is used for solving the inverse problem using the simplex search method (Nelder-Mead) [20] which does not require numerical or analytic gradients.

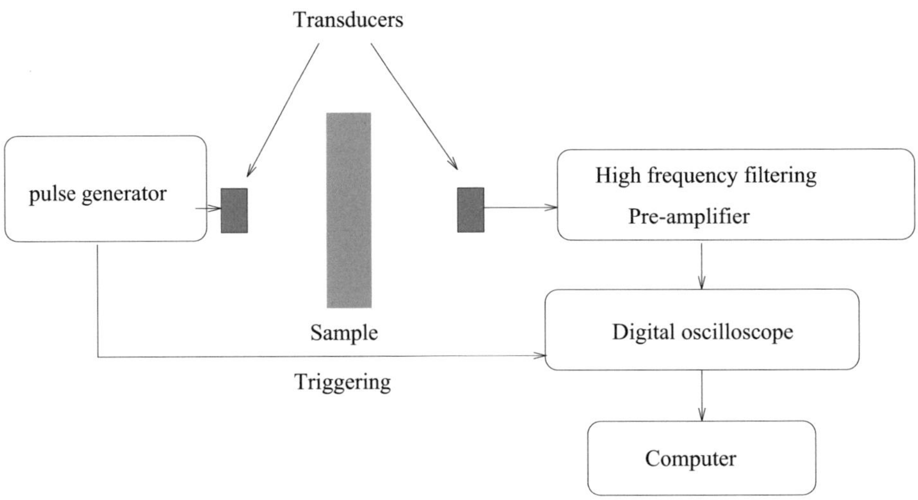

Figure 3. Experimental setup of the ultrasonic measurements.

http://dx.doi.org/10.5772/intechopen.72215

Consider a sample of plastic foam M1, of thicknesses 0.8 ± 0.01 cm. Sample M1 was characterized using classic methods [21–31] and gave the following physical parameters $\phi = 0.85 \pm 0.05$, $\alpha_\infty = 1.45 \pm 0.05$, $\Lambda = (30 \pm 1)\,\mu m$, and $\Lambda' = (60 \pm 3)\,\mu m$. **Figure 4** shows the experimental incident signal (dashed line) generated by the transducer and the experimental transmitted signal (solid line). After solving the inverse problem simultaneously for the porosity ϕ, tortuosity α_∞, and viscous and thermal characteristic lengths Λ and Λ', we find the following optimized values: $\phi = 0.87 \pm 0.01$, $\alpha_\infty = 1.45 \pm 0.01$, $\Lambda = (32.6 \pm 0.5)\,\mu m$, and $\Lambda' = (60 \pm 0.5)\,\mu m$. The values of the inverted parameters are close to those obtained by conventional methods [21–31]. We present in **Figures 5** and **6** the variation of the minimization function U with the porosity, tortuosity, viscous characteristic length, and the ratio between Λ' and Λ. In **Figure 7**, we show a comparison between an experimental transmitted signal and simulated transmitted signal for the optimized values of ϕ, α_∞, Λ, and Λ'. The difference between the two curves is small, which leads us to conclude that the optimized values of the physical parameters are correct.

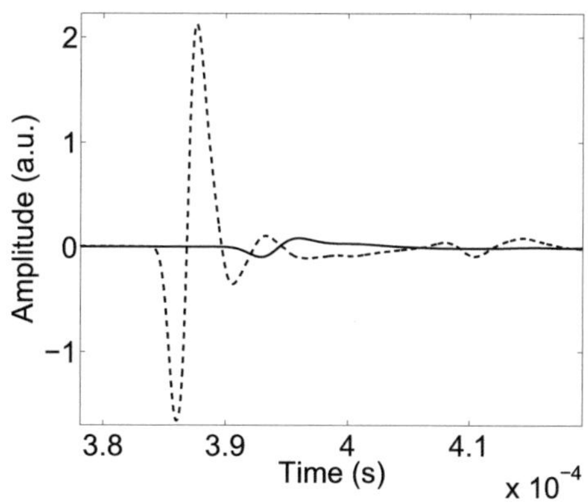

Figure 4. Experimental incident signal (solid line) and experimental transmitted signal (dashed line).

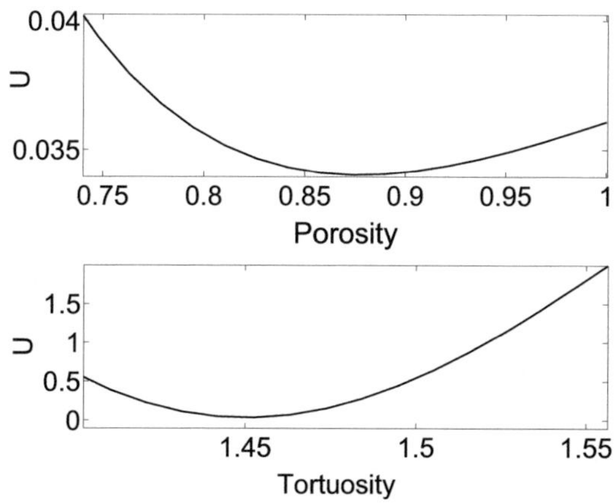

Figure 5. Variation of the minimization function U with porosity and tortuosity.

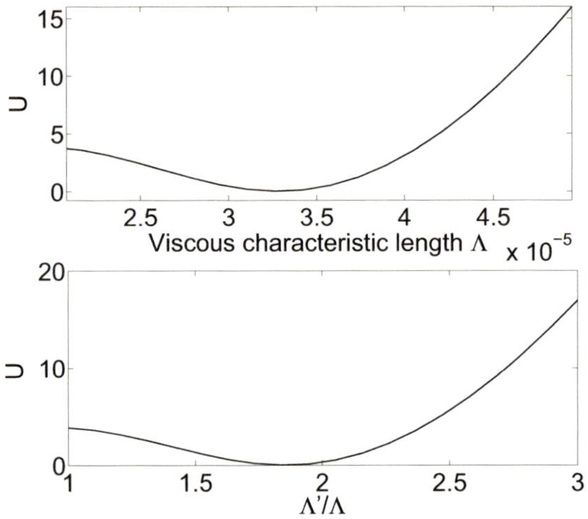

Figure 6. Variation of the cost function U with the viscous characteristic length Λ and the ratio Λ'/Λ.

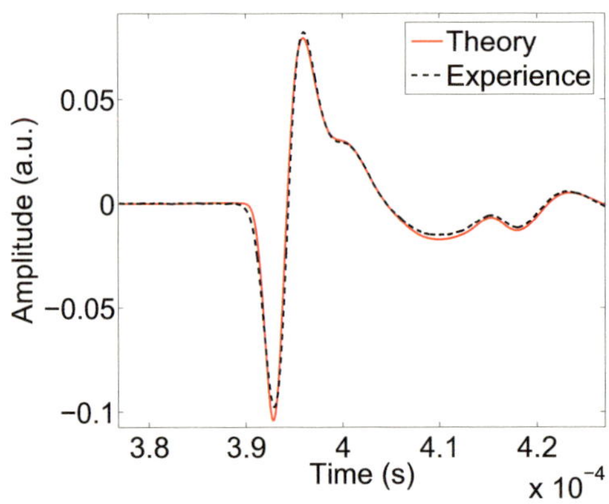

Figure 7. Comparison between the experimental transmitted signal (black dashed line) and the simulated transmitted signals (black line) using the reconstructed values of φ, α_∞, Λ, and Λ'.

3.2. Measuring flow resistivity of porous material via acoustic reflected waves at low-frequency domain

In the low-frequency domain, the viscous forces are important everywhere in all the fluid saturating the porous material. The thermal exchanges between fluid and structure are favored by the slowness of the cycle of expansion and compression in the material. The temperature of the frame is practically unchanged by the passage of the sound wave because of the high value of its specific heat: the frame acts as a thermostat; the isothermal compressibility is directly applicable. In this domain, the viscous skin thickness $\delta = (2\eta/\omega\rho_0)^{1/2}$ is much larger than the radius of the pores r

$$\frac{\delta}{r} \gg 1. \tag{34}$$

We consider the low-frequency approximations of the response factor $\alpha(\omega)$ and $\beta(\omega)$. When $\omega \to 0$, Eqs. (22) and (6), respectively, become

$$\alpha(\omega) = \frac{\sigma\phi}{i\omega\rho}, \tag{35}$$

$$\beta(\omega) = \gamma. \tag{36}$$

For a wave traveling along the direction ox, the generalized forms of the basic Eqs. (19) in the time domain are now

$$\sigma\phi V = -\frac{\partial p}{\partial x} \quad \text{and} \quad \frac{\gamma}{K_a}\frac{\partial p}{\partial t} = -\frac{\partial v}{\partial x} \tag{37}$$

where the Euler equation is reduced to Darcy's law which defines the static flow resistivity $\sigma = \eta/k_0$. The wave equation in time domain is given by

$$\frac{\partial^2 p}{\partial x^2} + \left(\frac{\sigma\phi\gamma}{K_a}\right)\frac{\partial p}{\partial t} = 0 \tag{38}$$

The fields which are varying in time, the pressure, the acoustic velocity, etc. follow a diffusion equation with the diffusion constant:

$$D = \frac{K_a}{\sigma\phi\gamma}. \tag{39}$$

The diffusion constant D is connected to Darcy's constant k_0 (called also the viscous permeability) by the relation

$$D = \frac{K_a k_0}{\eta\phi\gamma}, \tag{40}$$

where η is the fluid viscosity.

The expression of the reflection coefficient $R(z)$ in Laplace domain (put $z = j\omega$ for obtaining the frequency domain of $R(\omega)$), is given by [32]

$$R(z) = \frac{(1 - B^2 z)\sinh\left(L\sqrt{Dz}\right)}{2B\sqrt{z}\cosh\left(L\sqrt{Dz}\right) + (1 + B^2 z)\sinh\left(L\sqrt{Dz}\right)}, \tag{41}$$

The development of these expressions in exponential series leads to the reflection coefficient:

$$R(z) = \frac{1 - B\sqrt{z}}{1 + B\sqrt{z}}\sum_{n\geq 0}\left(\frac{1 - B\sqrt{z}}{1 + B\sqrt{z}}\right)^{2n}\left(\exp\left(-2nL\sqrt{Dz}\right) - \exp\left(-2(n+1)L\sqrt{Dz}\right)\right). \tag{42}$$

The multiple reflections in the material are taken into account in these expressions. As the attenuation is high in the porous materials, the multiple reflection effects are negligible. Let us consider the reflections at the interfaces $x = 0$ and $x = L$:

$$R(z) = \frac{1 - B\sqrt{z}}{1 + B\sqrt{z}}\left(1 - \frac{4B\sqrt{z}}{(1 + B\sqrt{z})^2}\exp\left(-2L\sqrt{Dz}\right)\right)$$
$$= \frac{1 - B\sqrt{z}}{1 + B\sqrt{z}} - \frac{4B\sqrt{z}(1 - B\sqrt{z})}{(1 + B\sqrt{z})^3}\exp\left(-2L\sqrt{Dz}\right) \tag{43}$$

The reflection scattering operator is calculated by taking the inverse Laplace transform of the reflection coefficient.

We infer [32] that

$$\mathcal{L}^{-1}\left[\frac{1 - B\sqrt{z}}{(1 + B\sqrt{z})}\right] = \mathcal{L}^{-1}\left[-1 + \frac{2}{B}\frac{1}{\sqrt{z} + 1/B}\right]$$
$$= -\delta(t) + \frac{2}{B\sqrt{\pi t}} - \frac{2}{B^2}\exp\left(t/B^2\right)\ \mathrm{erf}\left(\sqrt{t}/B\right), \tag{44}$$

where erf is the error function. By putting

$$g(z) = \frac{Bz - 1}{(1 + Bz)^3} = \frac{1}{B^2}\frac{z - 1/B}{(1/B + z)^3},$$

we obtain

$$\mathcal{L}^{-1}[g(z)] = f(t) = \frac{1}{B^2}\mathcal{L}^{-1}\left[\frac{z - 1/B}{(1/B + z)^3}\right] = \frac{1}{B^2}(t - t^2/B)\exp(-t/B).$$

Using the relation

$$\mathcal{L}^{-1}\left[\sqrt{z}g\left(\sqrt{(z)}\right)\right] = \frac{1}{2\sqrt{\pi}}\frac{1}{t^{3/2}}\int_0^\infty \exp\left(-\frac{u^2}{4t}\right)\left(\frac{u^2}{2t} - 1\right)f(u)du$$
$$= \frac{1}{2\sqrt{\pi}B^2}\frac{1}{t^{3/2}}\int_0^\infty \exp\left(-\frac{u^2}{4t}\right)\left(\frac{u^2}{2t} - 1\right)\left(u - \frac{u^2}{B}\right)\exp\left(-\frac{u}{B}\right)du,$$

which with the variable change $u/B = y$, yields

$$\mathcal{L}^{-1}\left[\frac{4B\sqrt{z}(B\sqrt{z} - 1)}{(1 + B\sqrt{z})^3}\right] = \frac{2}{B\sqrt{\pi}}\frac{1}{t^{3/2}}\int_0^\infty \exp\left(-\frac{u^2}{4t}\right)\left(\frac{u^2}{2t} - 1\right)\left(u - \frac{u^2}{B}\right)\exp\left(-\frac{u}{B}\right)du,$$
$$= \frac{2B}{\sqrt{\pi}}\frac{1}{t^{3/2}}\int_0^\infty \exp\left(-\frac{B^2y^2}{4t}\right)\left(\frac{y^2B^2}{2t} - 1\right)(y - y^2)\exp(-y)dy.$$
$$= k(t)$$

The reflection scattering operator is then given by

$$\tilde{R}(t) = (f(t) + k(t))*g(t) \tag{45}$$

3.2.1. *Acoustic parameter sensitivity*

Consider a sample of porous material having a physical parameters that correspond to quite common acoustic materials, as follows: thickness $L = 4\,\text{cm}$, porosity $\phi = 0.9$, flow resistivity $\sigma = 30000\,\text{N m}^{-4}\text{s}$, and radius of the pore $r = 70\,\mu m$. Let us study the sensitivity of the main parameters using numerical simulations of waves reflected by a porous material. Fifty percent variation is applied to the physical parameters (flow resistivity σ and porosity ϕ).

To obtain the simulated reflected waves, we use the incident signal given in **Figure 8** (dashed line). The result (reflected wave) is the wave given in the same figure (**Figure 8**) in solid line. The spectra of the two waves (incident and reflected) are given in **Figure 9**. From **Figure 8**, we can see that there is just an attenuation of the reflected wave without dispersion, since the two waves have the same spectral bandwidth (**Figure 9**). **Figure 8** shows the results obtained after reducing flow resistivity by 50% of its initial value. The wave in dashed line corresponds to the simulated reflected signal for $\sigma = 30000\,\text{N m}^{-4}\text{s}$ and the second one (solid line) to $\sigma = 15000\text{N m}^{-4}\text{s}$. The values of the porosity $\phi = 0.9$ and thickness $L = 4\,\text{cm}$ have been kept constant. When the flow resistivity is reduced, the amplitude of reflected wave decreases by 30% of its initial value. Physically, by reducing the flow resistivity, the medium is less resistive, since the viscous effects become less important in the porous material, and thus the amplitude of the reflected wave decreases. No change is observed in the reflected wave when reducing the porosity by 50% of its initial value. We can conclude that the porosity has no significant sensitivity in reflected mode.

For the propagation of transient signals at low frequency, a guide (pipe) [32], having a diameter of 5 cm and of length 50 m, is chosen. The pipe can be rolled without perturbations on experimental signals (the cutoff frequency of the tube $f_c \sim 4\,\text{kHz}$). The same microphone (Brüel & Kjær, 4190) is used for measuring the incident and reflected signals. Burst is

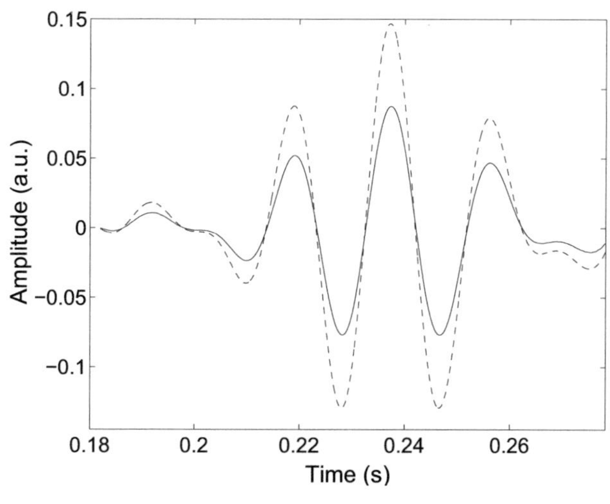

Figure 8. Incident signal (dashed line) and simulated reflected signal (solid line).

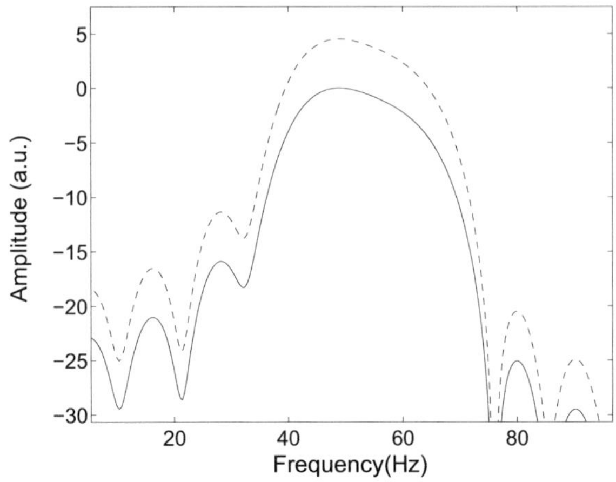

Figure 9. Spectrum of incident signal (dashed line) and spectrum of reflected signal (solid line).

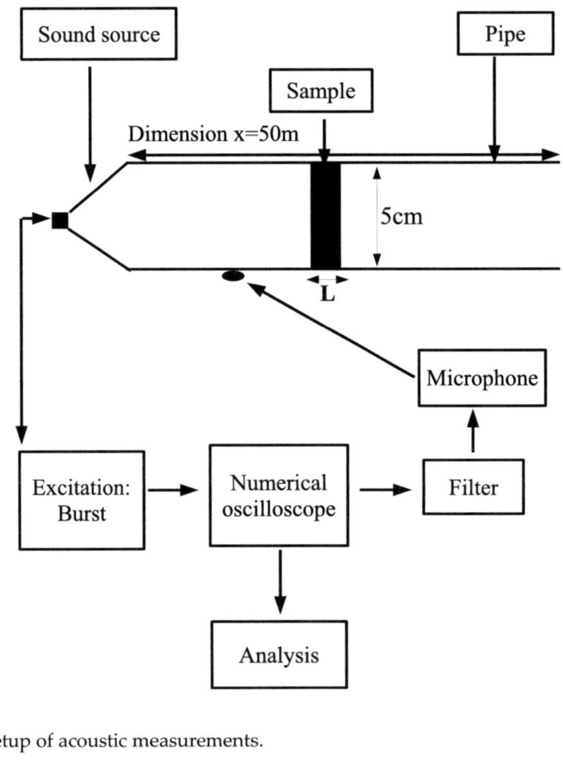

Figure 10. Experimental setup of acoustic measurements.

provided by synthesized function generator Stanford Research Systems model DS345-30 MHz. A sound source driver unit "Brand" constituted by loudspeaker Realistic 40-9000 is used. The incident signal is measured by putting a total reflector in the same position than the porous sample. The experimental setup is shown in **Figure 10**. Consider a cylindrical sample of plastic foam M1 of flow resistivity value $\sigma = 40000 \pm 6000\,\mathrm{Nm^{-4}s}$. This value is obtained using the method of Bies and Hansen [33]. The sample M1 has a diameter of 5 cm and a thickness of 3 cm. **Figure 11** shows the experimental incident wave (solid line) generated by the loudspeaker in the frequency bandwidth (35–75) Hz, and the experimental reflected signal (dashed line), with their spectra. There is no dispersion, since the two signals have practically the same bandwidth. The minimization of the function U gives the solution if the inverse problem:

$$U(\sigma) = \sum_{i=1}^{i=N} \left(p^{r}_{exp}(x, t_i) - p^{r}(x, t_i) \right)^2, \tag{46}$$

where $p^{r}_{exp}(x, t_i)_{i=1,2,\ldots N}$ and $p^{r}(x, t_i)_{i=1,2,\ldots N}$ represent the discrete set of values of the experimental reflected signal and of the simulated reflected signal, respectively. The optimized

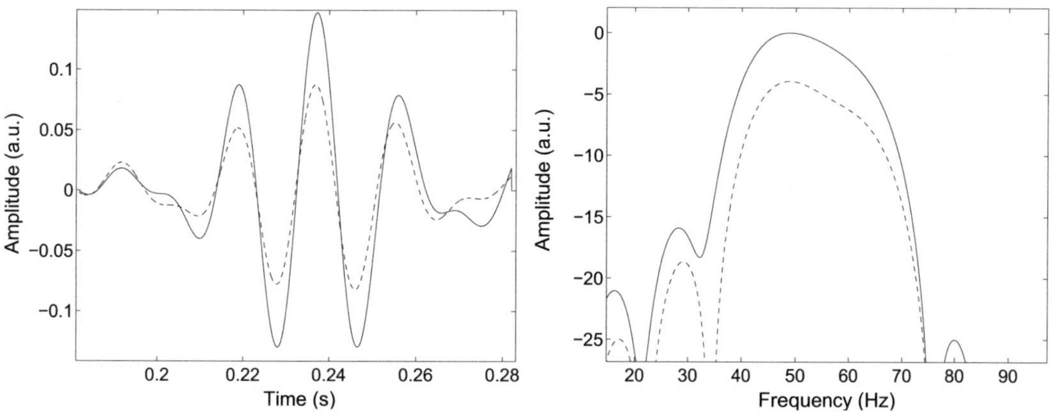

Figure 11. Experimental incident signal (solid line) and experimental reflected signal (dashed line), and their spectra, respectively.

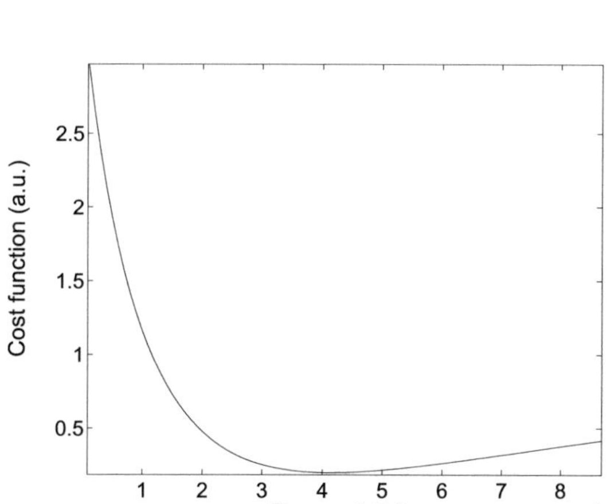

Figure 12. Variation of the minimization function U with flow resistivity σ.

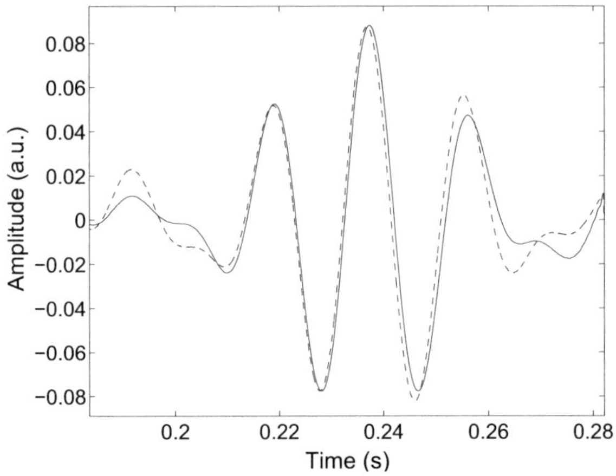

Figure 13. Comparison between experimental reflected signal (dashed line) and simulated reflected signal (solid line) for the sample M1.

value of $\sigma = 40500 \pm 2000\,\mathrm{Nm}^{-4}\mathrm{s}$ is obtained by solving the inverse problem. The variation of the minimization function U with the flow resistivity σ is given in **Figure 12**. A comparison between experiment and theory is given in **Figure 13**. The difference between theory and experiment is slight, which leads us to conclude that the optimized value of the flow resistivity is good.

This alternative acoustic method has the advantage of being simple and effective since it requires the use of only one microphone and therefore no calibration problem. In addition, this approach is different from conventional methods (Bies and Hansen [33]) that involve the use of fluid flow measurement techniques and pressure differences. The mathematical analysis of the reflected wave at low frequency is quite simple, because this wave is not propagative in the medium but simply diffusive (having the same frequency band with the incident signal). The wave reflected by the resistive materials has the advantage of being easily detectable experimentally compared to the transmitted wave.

4. Conclusion

Acoustic propagation in porous media involves a large number of physical parameters when the structure is elastic. This number is reduced when the structure is rigid, because the mechanical part does not intervene and thus remains only the acoustic part. The study of high and low frequencies separately solves the inverse problem and characterizes the porous materials in the domain of influence of the physical parameters. The proposed methods are simple and effective and allow an acoustic characterization of porous materials using transmitted or reflected experimental waves.

Author details

Zine El Abiddine Fellah[1]*, Mohamed Fellah[2], Claude Depollier[3], Erick Ogam[1] and Farid G. Mitri[4]

*Address all correspondence to: fellah@lma.cnrs-mrs.fr

1 LMA, CNRS, UPR 7051, Aix-Marseille Univ, Centrale Marseille, Marseille, France

2 Laboratoire de Physique Théorique, Faculté de Physique, USTHB, Bab-Ezzouar, Algeria

3 LUNAM Universite du Maine, UMR CNRS 6613 Laboratoire d'Acoustique de l'Universite du Maine UFR STS, Le Mans, France

4 Chevron, Area 52 - ETC, Santa Fe, New Mexico, United States

References

[1] Biot MA. The theory of propagation of elastic waves in fluid-saturated porous solid. I. Low frequency range. The Journal of the Acoustical Society of America. 1956;**28**:168

[2] Biot MA. The theory of propagation of elastic waves in fluid-saturated porous solid. I. Higher frequency range. The Journal of the Acoustical Society of America. 1956;**28**:179

[3] Johnson DL, Koplik J, Dashen R. Theory of dynamic permeability and tortuosity in fluid-saturated porous media. Journal of Fluid Mechanics. 1987;**176**:379-402

[4] Szabo TL. Time domain wave equations for lossy media obeying a frequency power law. The Journal of the Acoustical Society of America. 1994;**96**:491

[5] Szabo TL. Causal theories and data for acoustic attenuation obeying a frequency power law. Journal of the Acoustical Society of America. 1995;**97**:14

[6] Norton V, Novarini JC. Including dispersion and attenuation directly in time domain for wave propagation in isotropic media. The Journal of the Acoustical Society of America. 2003;**113**:3024

[7] Chen W, Holm S. Modified Szabo's wave equation models for lossy media obeying frequency power law. The Journal of the Acoustical Society of America. 2003;**113**:3024

[8] Fellah ZEA, Depollier C. Transient acoustic wave propagation in rigid porous media: A time domain approach. The Journal of the Acoustical Society of America. 2000;**107**(2): 683-688

[9] Fellah ZEA, Fellah M, Lauriks W, Depollier C, Angel Y, Chapelon JY. Solution in time domain of ultrasonic propagation equation in porous material. Wave Motion. 2003;**38**:151-163

[10] Fellah ZEA, Depollier C, Fellah M. Direct and inverse scattering problem in porous material having a rigid frame by fractional calculus based method. Journal of Sound and Vibration. 2001;**244**(2):359-366

[11] Allard JF. Propagation of Sound in Porous Media : Modeling Sound Absorbing Materials. London: Chapman and Hall; 1993

[12] Caputo M. Vibration of an infinite plate with a frequency dependent Q. The Journal of the Acoustical Society of America. 1976;**60**:634-639

[13] Bagley RL, Torvik PJ. On the fractional calculus model of viscoelastic behavior. Journal of Rheology. 1983;**30**:133-155

[14] Lafarge D, Lemarnier P, Allard JF, Tarnow V. Dynamic compressibility of air in porous structures at audible frequencies. Journal of the Acoustical Society of America. 1996;**102**:4

[15] Norris AN. On the viscodynamic operator in Biot's equations of poroelasticity. Journal of Wave-Material Interaction. 1986;**1**:365-380

[16] Samko SG, Kilbas AA, Marichev OI. Fractional Integrals and Derivatives Theory and Applications. Amsterdam: Gordon and Breach Publishers; 1993

[17] Fellah M, Fellah ZEA, Mitr FG, Ogam E, Depollier C. Transient ultrasound propagation in porous media using biot theory and fractional calculus: Application to human cancellous bone. The Journal of the Acoustical Society of America. 2013;**133**(4):683-688

[18] Allard JF, Champoux Y. New empirical equations for sound propagation in rigid frame fibrous materials. Journal of the Acoustical Society of America. 1992;**91**:3346-3353

[19] Fellah ZEA, Sadouki M, Fellah M, Mitri FG, Ogam E, Depollier C. Simultaneous determination of porosity, tortuosity, viscous and thermal characteristic lengths of rigid porous materials. Journal of Applied Physics. 2013;**114**:204902-204905

[20] Lagarias JC, Reeds JA, Wright MH, Wright PE. Convergence properties of the Nelder–mead simplex method in low dimensions. SIAM Journal on Optimization. 1998;**9**:112Ű147

[21] Leclaire P, Kelders L, Lauriks W, Glorieux C, Thoen J. Determination of the viscous characteristic length in air-filled porous materials by ultrasonic attenuation measurements. The Journal of the Acoustical Society of America. 1996;**99**:1944

[22] Ayrault C, Moussatov A, Castagnède B, Lafarge D. Ultrasonic characterization of plastic foams via measurements with static pressure variations. Applied Physics Letters. 1999;**74**:3224

[23] Moussatov A, Ayrault C, Castagnède B. Porous material characterization ultrasonic method for estimation of tortuosity and characteristic length using a barometric chamber. Ultrasonics. 2001;**39**:195

[24] Leclaire P, Kelders L, Lauriks W, Brown NR, Melon M, Castagnède B. Determination of viscous and thermal characteristic lengths of plastic foams by ultrasonic measurements in helium and air. Journal of Applied Physics. 1996;**80**:2009

[25] Fellah ZEA, Depollier C, Berger S, Lauriks W, Trompette P, Chapelon JY. Determination of transport parameters in air saturated porous materials via ultrasonic reflected waves. The Journal of thc Acoustical Society of America. 2003;**113**(5):2561-2569

[26] Fellah ZEA, Berger S, Lauriks W, Depollier C, Chapelon JY. Inverse problem in air-saturated porous media via reflected waves. Review of Scientific Instruments. 2003;**74**(5):2871

[27] Fellah ZEA, Berger S, Lauriks W, Depollier C, Aristégui C, Chapelon JY. Measuring the porosity and the tortuosity of porous materials via reflected waves at oblique incidence. The Journal of the Acoustical Society of America. 2003;**113**(5):2424

[28] Fellah ZEA, Berger S, Lauriks W, Depollier C, Trompette P, Chapelon JY. Ultrasonic measuring of the porosity and tortuosity of air- saturated random packings of beads. Journal of Applied Physics. 2003;**93**:9352

[29] Fellah ZEA, Mitri FG, Depollier D, Berger S, Lauriks W, Chapelon JY. Characterization of porous materials having a rigid frame via reflected waves. Journal of Applied Physics. 2003;**94**:7914

[30] Fellah ZEA, Berger S, Lauriks W, Depollier C, Fellah M. Measurement of the porosity of porous materials having a rigid frame via reflected waves : A time domain analysis with fractional derivatives. Journal of Applied Physics. 2003;**93**:296

[31] Fellah ZEA, Mitri FG, Fellah M, Ogam E, Depollier C. Ultrasonic characterization of porous absorbing materials: Inverse problem. Journal of Sound and Vibration. 2007;**302**:746-759

[32] Sebaa N, Fellah ZEA, Fellah M, Lauriks W, Depollier C. Measuring flow resistivity of porous material via acoustic reflected waves. Journal of Applied Physics. 2005;**98**:084901

[33] Bies DA, Hansen CH. Flow resistance information for acoustical design. Applied Acoustics. 1980;**13**:357-391

A Novel Idea of Coherent Acoustic Wave-Induced Atmospheric Refractivity Fluctuation and Its Applications

Shuhong Gong, Yu Liu, Muyu Hou and Lixin Guo

Additional information is available at the end of the chapter

http://dx.doi.org/10.5772/intechopen.70996

Abstract

The physical mechanism of generating the lasting tropospheric refractivity fluctuation with a stable array-distributed structure by coherent acoustic waves is investigated. An example of the quantitative calculation of atmospheric refractive index is given and analyzed. Based on the theory of electromagnetic wave propagation and scattering in the troposphere, the feasibility to purposefully affect radio wave propagation is qualitatively demonstrated by the experiment of the coherent acoustic source-induced laser interference fringe change. The potential application aspects of synthetically controlling the radio wave propagation by the artificial refractivity fluctuation structure are preliminarily proposed. This chapter will promote the development of the coherent acoustic wave-induced tropospheric refractivity fluctuation, and it has the important theoretical significance and potential application value to purposely apply the positive or negative effects on radio wave propagation.

Keywords: troposphere, atmospheric refractive index, coherent acoustic waves radio wave propagation, positive or negative effects

1. Introduction

Radio wave propagation is an important part of wireless systems. The investigations into various environment channel media with different physical properties, different dielectric properties, and spatiotemporal structures are inevitable to research radio wave propagation. There are two types of influences induced by the environment media on the radio wave propagation: the one is to realize particular propagation mode and advanced wireless technology, such as to realize the over-the-horizon communication or detection by atmospheric turbulence scattering and the multiple-input multiple-output wireless system by multipath propagation, which is seen as the positive influence; the other is to restrict radio signal propagation to be received or detected. For instance, the propagation effects of attenuation, depolarization, scintillation, and atmospheric noise are restrictive effects on communication systems, which are seen as the negative influence. It is the goal of the research on radio wave propagation to investigate the propagation environment media-caused positive influence and negative one and to apply them to the two aspects, based on the precondition of sufficiently mastering and controlling the physical and dielectric properties of propagation environment media. On the one hand, it makes the performances of an expected wireless electronic system perfectly match with its wireless channel, by forecasting, modifying, and applying the positive effects. On the other hand, it destroys the original perfect match between the hostile wireless electronic system and its wireless channel during wireless system countermeasures, by forecasting, creating, and controlling the negative effects.

The idea of coherent acoustic wave-induced atmospheric refractivity fluctuation is to control the characteristic parameters of radio wave, such as amplitude, phase, propagation direction, and polarization, by the propagation effects caused by the artificial atmospheric refractivity fluctuation. And, the final aim is to purposefully apply the positive effects or the negative effects. Note that the idea of coherent acoustic wave-induced atmospheric refractivity fluctuation and its application is a new field and is in its infant stage.

The idea of changing the tropospheric refractivity by the disturbance of acoustic wave is first proposed in [1], in which A. Tonning pointed out the viewpoint that an acoustic wave propagating in the troposphere can cause the atmospheric refractivity fluctuation and theoretically analyzed the rationality of the viewpoint. In the next 50 years, radio acoustic sounding system (RASS) began to be established to detect atmospheric temperature, wind profile, and turbulence in the lower troposphere [2–4]. The first RASS system to measure atmospheric temperature profile was born at Stanford University [5]. By the 1990s, this technique was adopted to solve the problems of measuring the temperature in indoor environment [6–9], also employed to detect the wake vortex of aircrafts [10, 11], and even explored to track and detect the taggant for the soldier identification friend-or-foe application [12]. Currently, RASS can detect the atmospheric parameters below 20 km [13]. The RASS history was summarized, and the applications of RASS in the detection of turbulence were analyzed in [14]. The applications and limitations of RASS system are analyzed in [15]. Afterward, the broadband acoustic pulse technology [16] and imaging techniques [17] are applied to RASS system.

The idea of coherent acoustic wave-induced tropospheric refractivity fluctuation mentioned in this chapter is fundamentally different from RASS. The main idea of RASS is to obtain the parameters of atmospheric physical properties based on the relationship between acoustic propagation velocity and atmospheric physical property parameters by tracking the velocity of acoustic wave front which can be seen as an artificial refractivity irregularity. However, the main idea, in this chapter, is to purposely apply the positive influences and negative ones induced by artificial array-distributed refractivity irregularity, which is stable, lasting, and controllable because the coherent acoustic source is used. The viewpoint to purposely affect radio wave propagation by a coherent acoustic source is proposed for the first time in [18]. In this chapter, the physical mechanism of generating the lasting tropospheric refractivity fluctuation with stable array-distributed structure by coherent acoustic waves is elaborated in Section 2, the distribution of the artificial atmospheric refractive index is quantitatively calculated, and the feasibility to purposefully affect radio wave propagation is qualitatively demonstrated by the experiment of the coherent acoustic source-induced laser interference fringe change in Section 3. The potential application aspects of synthetically controlling the radio wave propagation by the artificial refractivity fluctuation structure are preliminarily proposed in Section 4; further investigations in the future are listed in Section 5. This chapter will promote the development of the coherent acoustic wave-induced tropospheric refractivity fluctuation.

2. The mechanism of controlling the atmospheric refractivity fluctuation by coherent acoustic waves

The real part N of the atmospheric refraction index in the radio band is nearly independent of frequency. N depends on the atmospheric pressure P in hPa, the absolute temperature T in K, and the water vapor pressure e in hPa, and the relation among them is [19]

$$N = 77.6\frac{P}{T} + 373256\frac{e}{T^2} \tag{1}$$

Operating the difference algorithm on Eq. (1), the relation among ΔN, ΔT, ΔP, and Δe is given as

$$\Delta N = -\left[77.6PT^{-2} + 746512eT^{-3}\right]\Delta T + \left(373265eT^{-2}\right)\Delta e + \left(77.6T^{-1}\right)\Delta P \tag{2}$$

Eq. (2) shows that ΔN is closely related to ΔT, ΔP, and Δe. The refractivity fluctuation ΔN can be steered by controlling one or more parameters among ΔT, ΔP, and Δe according to Eq. (2).

Acoustic wave movement follows the three basic laws: Newton's second law of motion, law of conservation of mass, and thermodynamic equation of state [20]. The wave function of an acoustic wave can be derived from the three basic laws and is written as [21]

$$\nabla^2 p = \frac{1}{c_0}\frac{\partial^2 p}{\partial t^2} \tag{3}$$

where p is the instantaneous acoustic pressure and c_0 denotes the acoustic propagation velocity in the atmosphere. p and c_0 can be expressed as

$$p = \Delta P = P - P_0 \tag{4}$$

$$c_0 = 331.6 + 0.6t_t \tag{5}$$

In Eq. (4), P_0 and P are the atmospheric pressure in Pa before and after the disturbance by an acoustic wave, respectively. And t_t in Eq. (5) is the atmospheric temperature in °C. Eq. (3) shows that when an acoustic wave passes through the atmospheric medium, the additional periodical variational pressure, that is, acoustic pressure $p = \Delta P$, will be exerted upon the medium based on the original atmospheric pressure. It is rational to neglect the temperature change induced by the additional periodical variational pressure; therefore $\Delta T=0$. According to [18], ΔP and Δe satisfy the formula $\Delta e/e=\Delta P/P$. Therefore, Eq. (2) can be deduced as

$$\Delta N=\left(77.6T^{-1}+373265e^{2}T^{-2}P^{-1}\right)\cdot p \tag{6}$$

It is shown in Eq. (6) that the refractivity fluctuation ΔN will present a specific spatial and temporal distributions along with the acoustic pressure p.

The solutions to Eq. (3) for the cases of the plane wave, the cylindrical wave, and the spherical wave can be, respectively, formulated as

$$p=p_{A}e^{j\left(\omega t-kr_{p}\right)} \tag{7}$$

$$p=\frac{p_{A}}{\sqrt{r_{c}-l_{0}}}e^{j[\omega t-k(r_{c}-l_{0})]} \tag{8}$$

$$p=\frac{p_{A}}{r_{s}-r_{0}}e^{j[\omega t-k(r_{s}-r_{0})]} \tag{9}$$

In Eqs. (7) and (9), p denotes the instantaneous value of the acoustic pressure at a point in their acoustic pressure field, r_p represents the distance between a planar acoustic source to a point in the acoustic pressure field, r_s represents the distance between the center axis of a cylinder-surface acoustic source to a point in the acoustic pressure field, r_0 represents the distance between the center of a spherical-surface acoustic source and a point in the acoustic pressure field, r_c is the diameter of the spherical-surface acoustic source, l_0 is the length of the cylindrical-surface acoustic source, and p_A is the acoustic pressure on the surface of the above acoustic sources. ω and k denote the angular frequency and wave number of the acoustic waves, respectively.

p_A for the three types of acoustic sources mentioned above can be, respectively, expressed as

$$p_{A}=\begin{cases}\left[2W\rho_{0}c_{0}S^{-1}\right]^{1/2} & \text{For plane acoustic source}\\ \left[2W\rho_{0}c_{0}(2\pi r_{0}l_{0})^{-1}\right]^{1/2} & \text{For cylindrical-surface acoustic source}\\ \left[2W\rho_{0}c_{0}\left(4\pi r_{0}^{2}\right)^{-1}\right]^{1/2} & \text{For spherical-surface acoustic source}\end{cases} \tag{10}$$

where W is the acoustic wave power radiated by the acoustic sources, S is the area of the plane acoustic source, and ρ_0 is the density of air.

In other words, when a spherical-surface acoustic wave passes through the homogeneous atmosphere medium, the distribution of acoustic pressure p or atmospheric refractivity fluctuation ΔN at a certain moment in a plane including the point acoustic source shown in **Figure 1** is similar to the waveform of a mechanical wave in water surface shown in **Figure 2**. Correspondingly, the refractivity fluctuation ΔN will present a similar spatial distribution following

Figure 1. The spatial distribution of p or ΔN when a spherical-surface acoustic wave travel through the homogeneous medium.

Figure 2. The waveform of a water surface wave caused by a single source.

Eq. (6). It is by means of the reflection effect of the acoustic wave front with ΔN that the purpose of tracking the propagation speed of acoustic wave is realized in RASS.

The principle of independent propagation and superposition of waves indicates that if two coherent waves meet, the interference phenomenon can be formed. **Figure 3** shows the waveform when two coherent water surface waves meet. **Figure 4** shows the simulation of the distribution of p or ΔN when two coherent acoustic waves meet.

According to the wave interference theory, interference pattern is closely related to coherent wave frequency, phase, acoustic source structure, and so on. For any type of coherent acoustic source with array structure, the acoustic pressure at point $\vec{r}$ in the interference area can be expressed as

Figure 3. The waveform when two coherent water surface waves meet.

Figure 4. The distribution of p or ΔN when two coherent acoustic waves meet.

$$p\left(\vec{r}\right) = p_0\left(\vec{r}\right)|S| \tag{11}$$

where $p_0\left(\vec{r}\right)$ gives the acoustic pressure of single elements of the acoustic source array and $|S|$ is the array factor, which presents the interference process. The array factor of the uniform spherical-surface acoustic source planar array shown in **Figure 5** and the uniform cylindrical-surface acoustic source linear array shown in **Figure 6** are, respectively, given as [18]

$$|S(\theta,\phi)| = \left| \frac{\sin\left[N_x \frac{\pi b_x \sin\theta\cos\phi}{\lambda} + \beta_x\right]}{\sin\left[\frac{\pi b_x \sin\theta\cos\phi}{\lambda} + \beta_x\right]} \cdot \frac{\sin\left[N_y \frac{\pi b_y \sin\theta\sin\phi}{\lambda} + \beta_y\right]}{\sin\left[\frac{\pi b_y \sin\theta\sin\phi}{\lambda} + \beta_y\right]} \right| \tag{12}$$

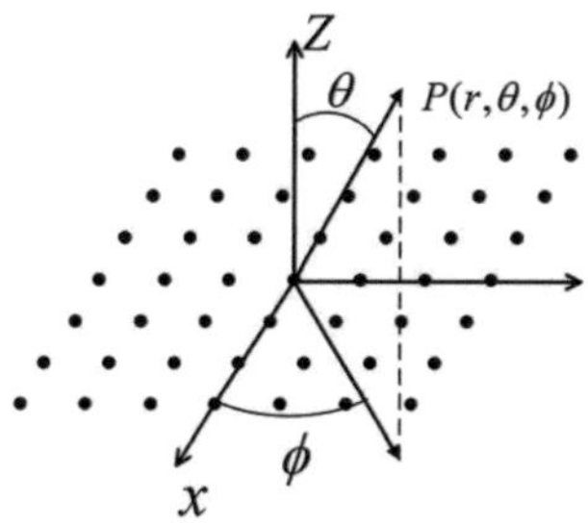

Figure 5. The sketch of a uniform spherical-surface acoustic source planar array.

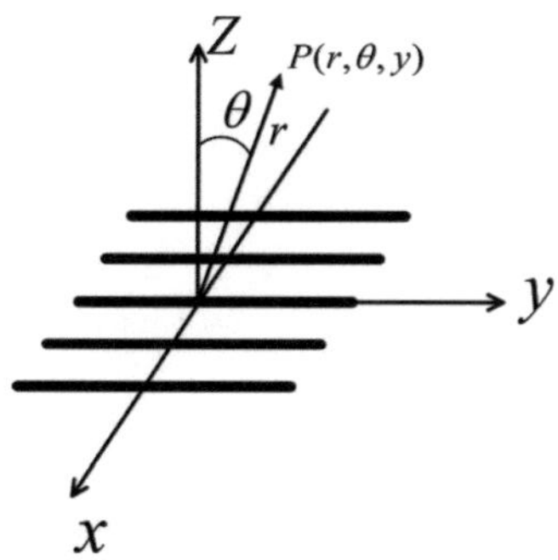

Figure 6. The sketch of a uniform cylinder-surface acoustic source linear array.

$$|S(\theta)| = \left| \frac{\sin\left[N\frac{b\sin\theta}{\lambda} + \beta\right]}{\sin\left[\frac{b\sin\theta}{\lambda} + \beta\right]} \right| \tag{13}$$

In Eq. (12), N_x and N_y are, respectively, the number of array elements along the x-axis and y-axis, and b_x and b_y are, respectively, the inter-element spacing along the x- and y-axes in **Figure 5**. In Eq. (13), N and b denote, respectively, the element number and the interval in **Figure 6**. λ is the acoustic wavelength. β_x and β_y are the harmonic vibration phase differences between the adjacent elements along x-axis and y-axis in **Figure 5**, respectively. β is the harmonic vibration phase difference between the adjacent elements in **Figure 6**. **Figures 7** and **8** are the simulating examples according to Eq. (12) and Eq. (13).

Therefore, the lasting artificial atmospheric refractivity irregularities with a stable, controllable, and subtle array structure in a specific space can be generated by setting the geometric structure of acoustic source and adjusting the frequency, amplitude, and phrase of acoustic waves. In [18], the coherent acoustic source shown in **Figure 9** and the short scattering communication link shown in **Figure 10** are used in the experiment. The experimental results as shown in **Figures 11–18** are observed, which further verify the feasibility of perturbation in the tropospheric atmospheric refractivity by coherent acoustic wave.

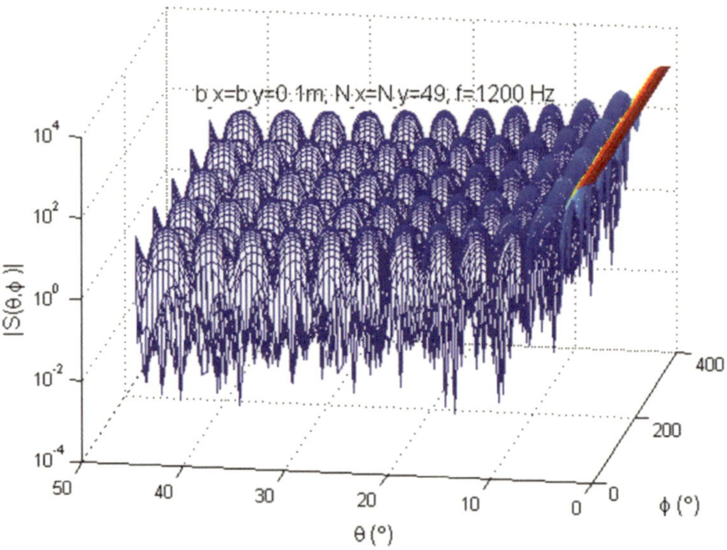

Figure 7. The simulation example of the array factor of the acoustic antenna array in **Figure 5.**

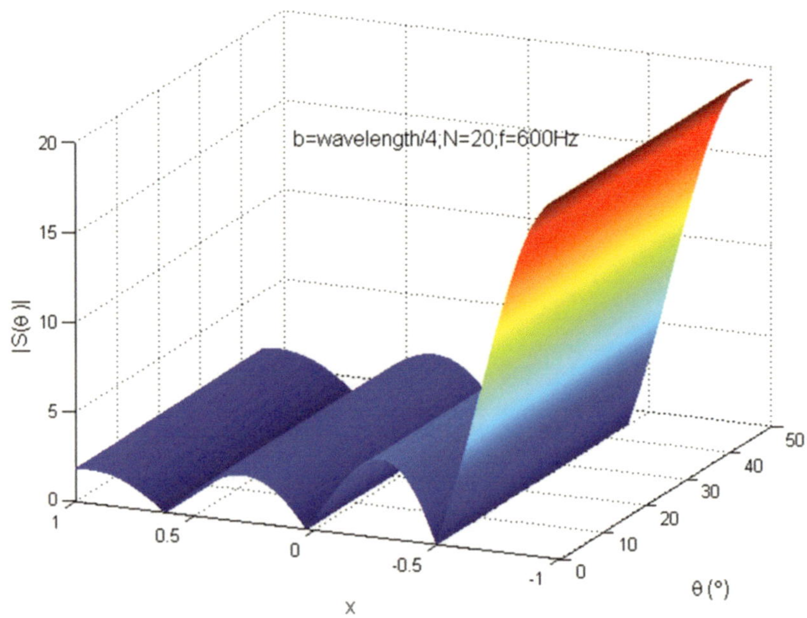

Figure 8. The simulation example of the array factor of the acoustic antenna array in **Figure 6.**

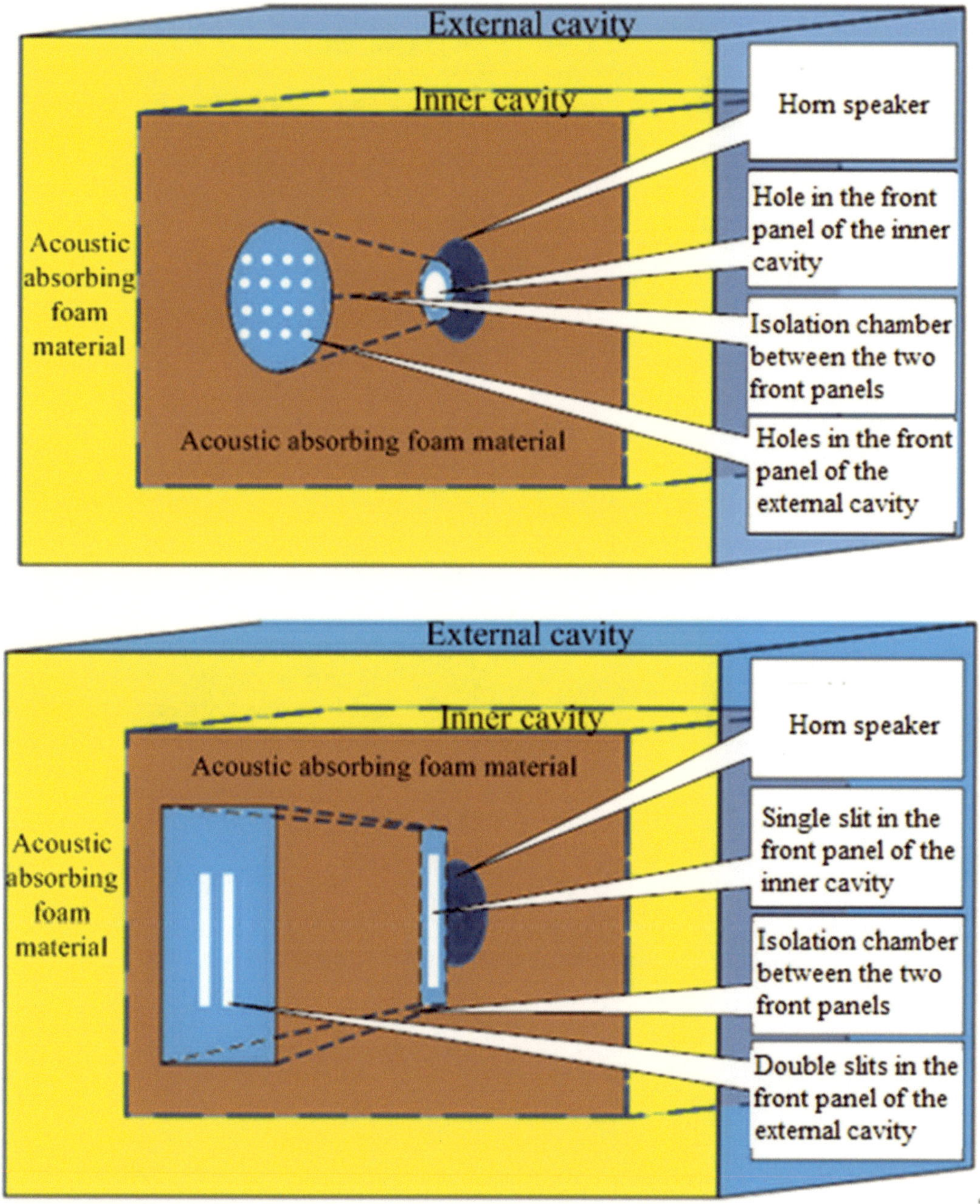

Figure 9. The sketch of the coherent acoustic source adopted in [18].

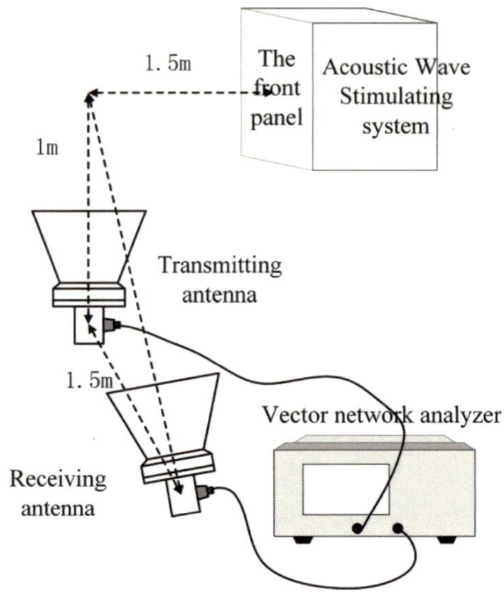

Figure 10. The sketch of the radio link adopted in the testing experiment in [18].

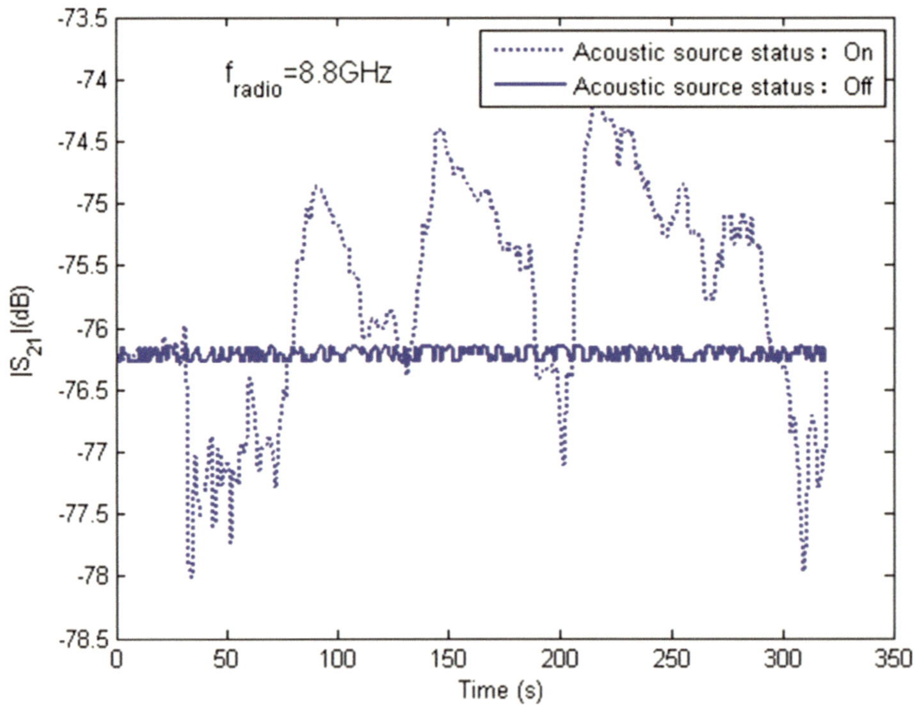

Figure 11. The testing result of 8.8 GHz stimulated by the "double-slit" acoustic source.

http://dx.doi.org/10.5772/intechopen.70996

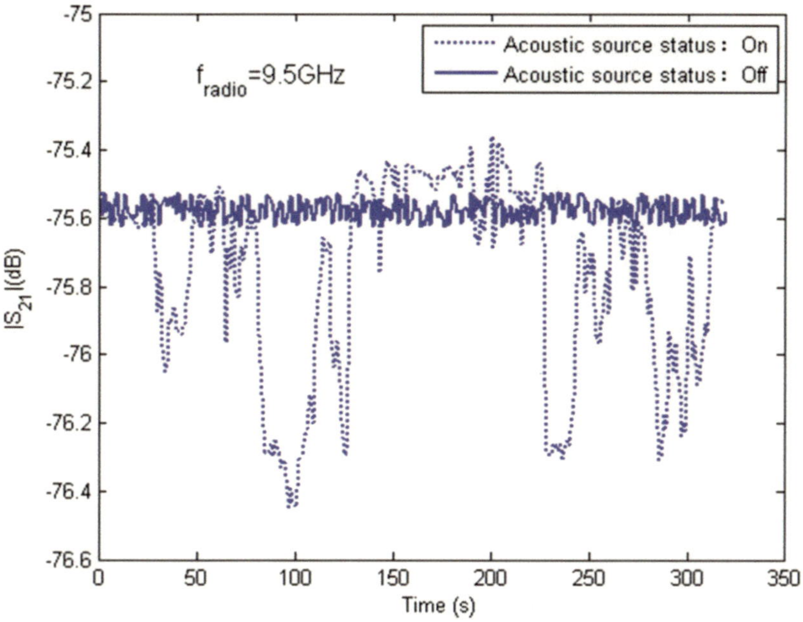

Figure 12. The testing result of 9.5 GHz stimulated by the "double-slit" acoustic source.

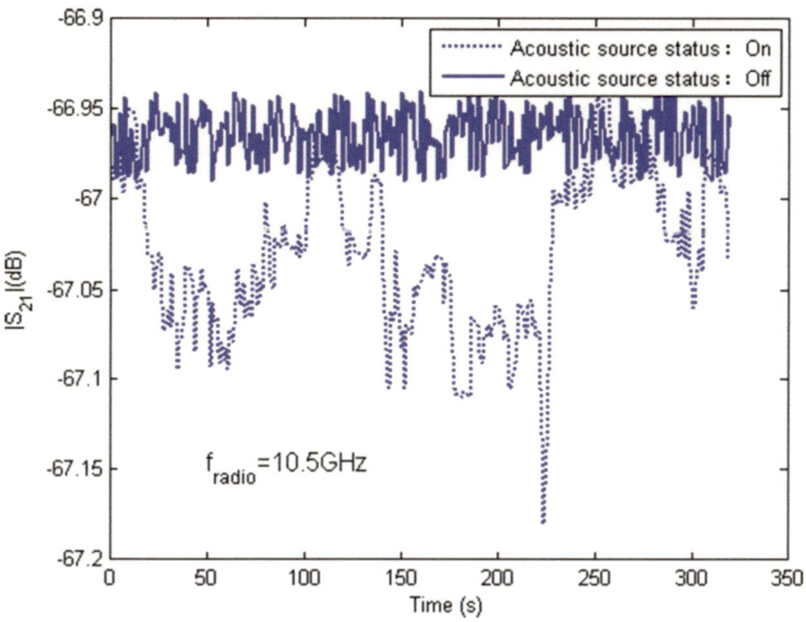

Figure 13. The testing result of 10.5 GHz stimulated by the "double-slit" acoustic source.

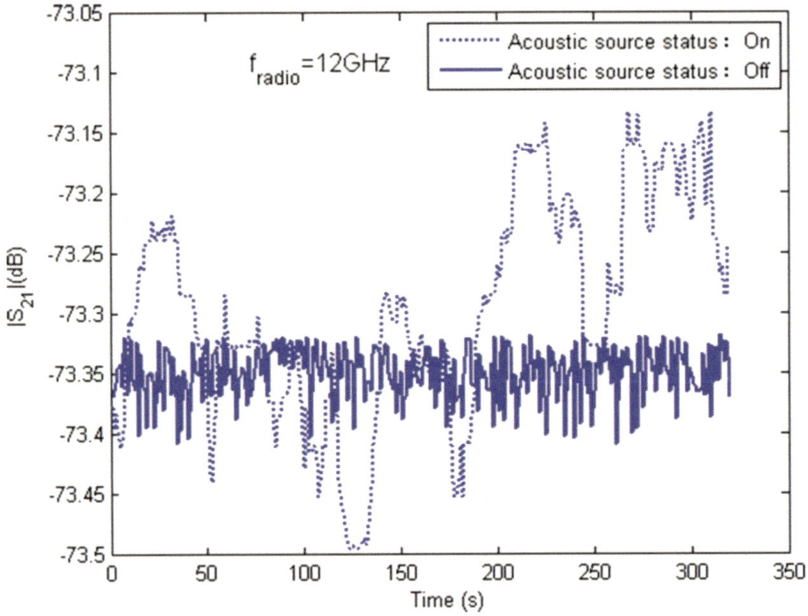

Figure 14. The testing result of 12 GHz stimulated by the "double-slit" acoustic source.

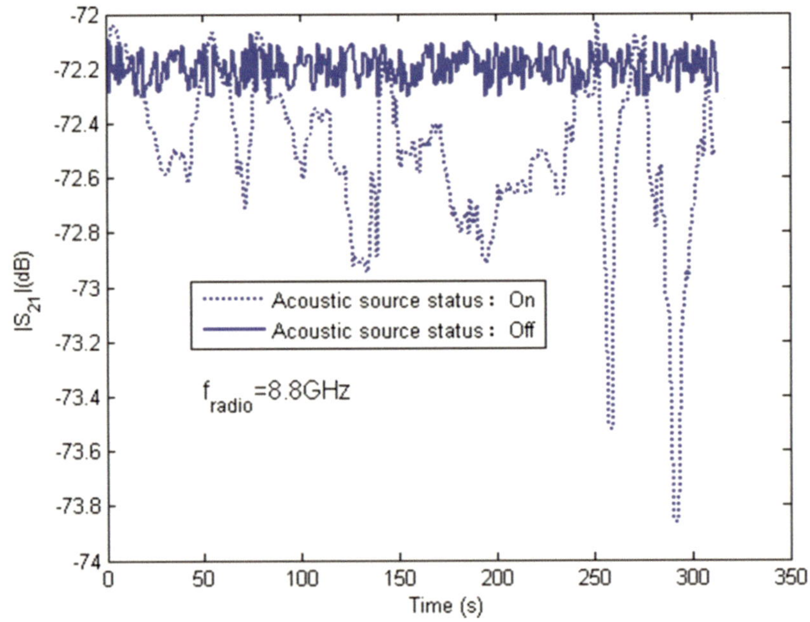

Figure 15. The testing result of 8.8 GHz stimulated by the "7 × 7 hole" acoustic source.

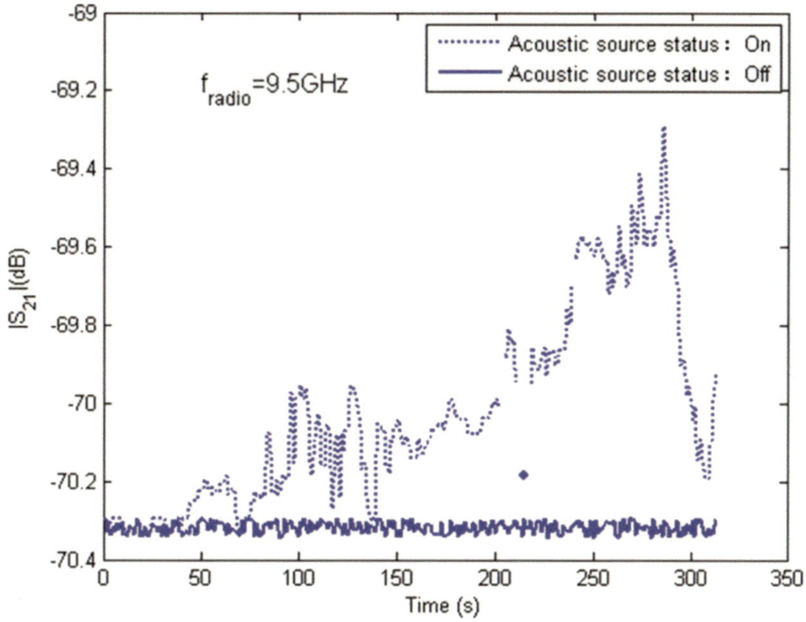

Figure 16. The testing result of 9.5 GHz stimulated by the "7 × 7 hole" acoustic source.

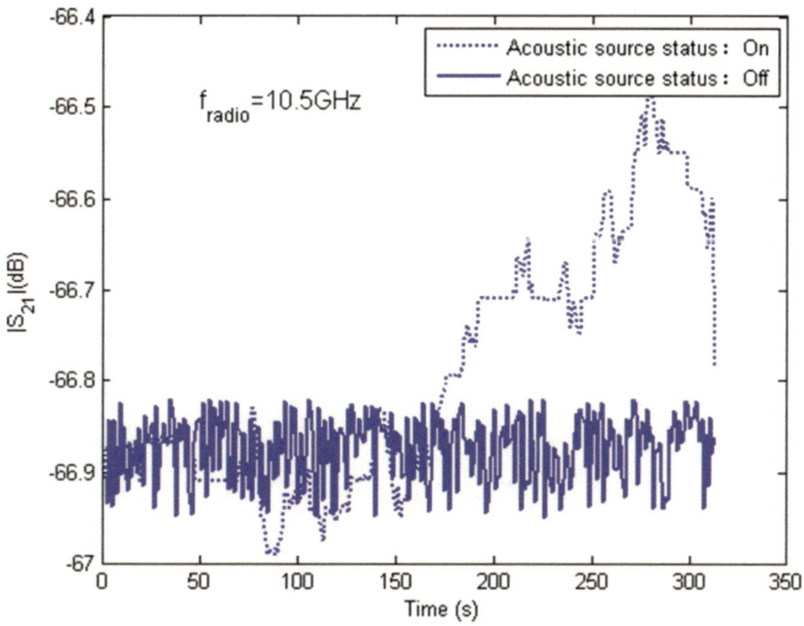

Figure 17. The testing result of 10.5 GHz stimulated by the "7 × 7 hole" acoustic source.

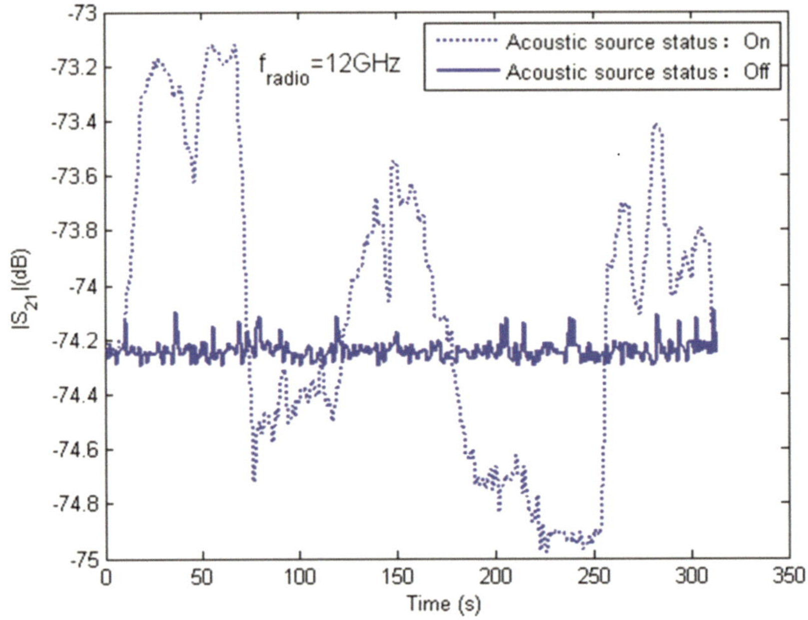

Figure 18. The testing result of 12 GHz stimulated by the "7 × 7 hole" acoustic source.

3. The interaction between a radio wave and the artificial refractivity fluctuation

According to [18], the spatial distribution of the acoustic pressure caused by the uniform spherical-surface acoustic source planar array shown in **Figure 5** is

$$p = \frac{p_A}{r - r_0} e^{-\alpha(r-r_0)} e^{j[\omega t - k(r-r_0)]} |S(\theta, \phi)| \tag{14}$$

where α denotes the attenuation coefficient of acoustic wave and $|S(\theta, \phi)|$ is the array factor of Eq. (12). The instantaneous ΔN_{C_I} and the effective ΔN_{C_E} of the atmospheric refractivity fluctuation ΔN at a space point are given by

$$\Delta N_{C\_I} = \left(77.6T^{-1} + 373265e^2T^{-2}P^{-1}\right) \cdot 10^{-2} \cdot p \tag{15}$$

$$\Delta N_{C\_E} = 0.707 \cdot \left(77.6T^{-1} + 373265e^2T^{-2}P^{-1}\right) \cdot 10^{-2} \cdot |p| \tag{16}$$

where p is the instantaneous acoustic pressure in Pa and $|p|$ denotes the amplitude of the acoustic pressure harmonically changing over time in Pa. The effective acoustic pressure can be given as $0.707|p|$.

Figures 19 and **20** show the spatial distributions of atmospheric refractivity fluctuation caused by the 49 × 49 hole acoustic source. In the calculation, the attenuation coefficient of acoustic wave α and the diameter of array element are ignored. The atmospheric pressure P is 1013.25 hPa, the

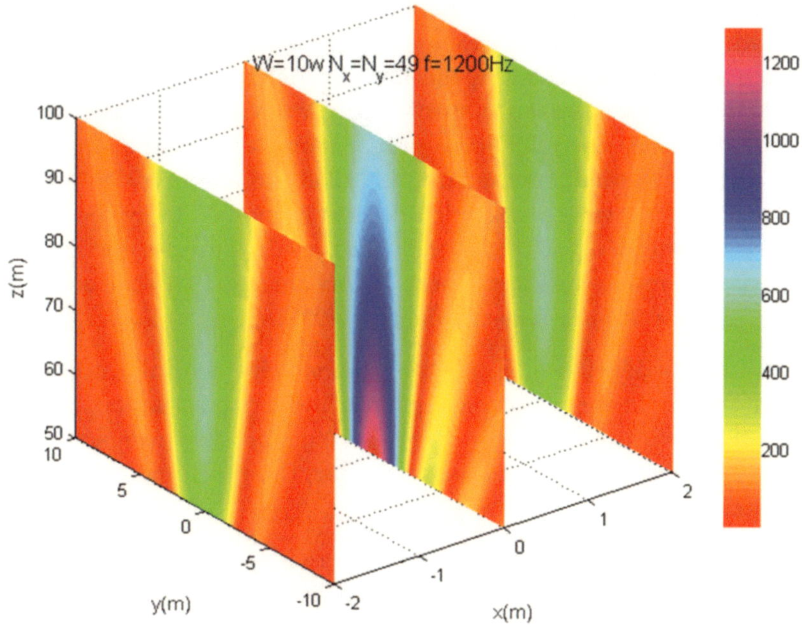

Figure 19. The spatial distribution of atmospheric refractivity parallel to the z-axis.

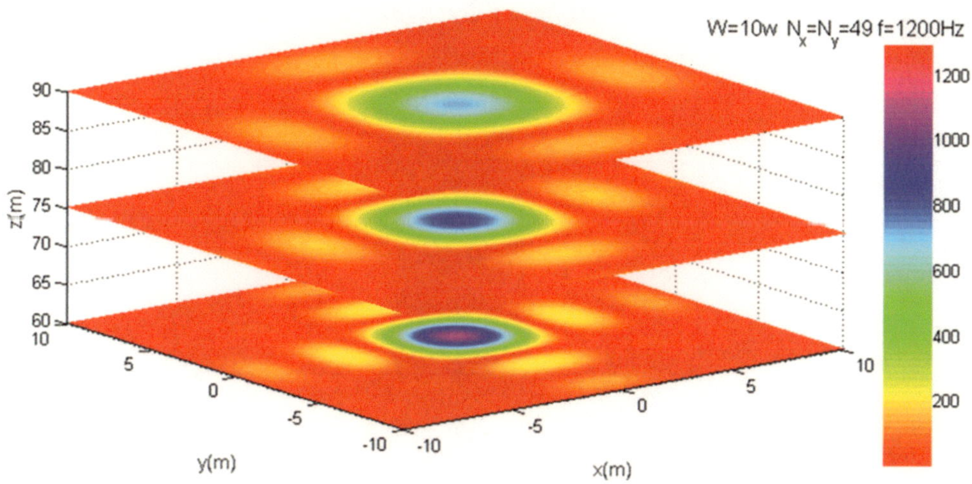

Figure 20. The spatial distribution of atmospheric refractivity parallel to the X-Y plane.

temperature T is 288.15 K, the water vapor pressure $e$is 10.02 hPa, and the power of each acoustic source element is 10 W.

As can be seen from **Figures 19** and **20**, when the coherent acoustic wave travels through the tropospheric medium, there will be a periodical atmospheric refractivity fluctuation. The mechanism the artificial dielectric irregularities impacting the radio wave propagation is similar to that the atmospheric turbulence impacting the radio wave propagation. The atmospheric refractivity fluctuation caused by the atmospheric turbulence is stochastic distributed, and that caused by acoustic wave shows spatial periodic distribution. Therefore, the influences of atmospheric turbulence on radio wave propagation need to be analyzed by the wave propagation theory in the random medium [22], and the effects of the artificial dielectric irregularities on the radio wave propagation need to be studied using the theory of electromagnetic scattering, reflection, and refraction by the periodic structure medium. Bragg volume scattering theory and Fresnel volume scattering theory are mainly used to analyze the impacts of artificial irregularities on radio wave propagation [23–26].

Our research group is devoted to quantitatively analyzing the influences of the artificial irregularities on the amplitude, phase, propagation direction, polarization, and other parameters of an electromagnetic wave. The related exploration is possible to implement leap-forward development in the fields of electronic countermeasure, communications, radar detection, and other wireless technologies.

An experiment aims to influence Michelson interference fringes by the coherent acoustic wave-induced atmospheric refractivity fluctuation was carried out, and it qualitatively verifies the feasibility to purposefully change the phase of the electromagnetic wave by the artificial refractivity fluctuation. As shown in **Figure 21**, a light emitted by laser emits hits the beam splitter G1, which is partially reflective. One part of the light denoted by 1 is reflected, while another part of the light denoted by 2 is transmitted through G1. The path of light 1 passes through the region disturbed by the artificial refractivity fluctuation, while the light 2 travels

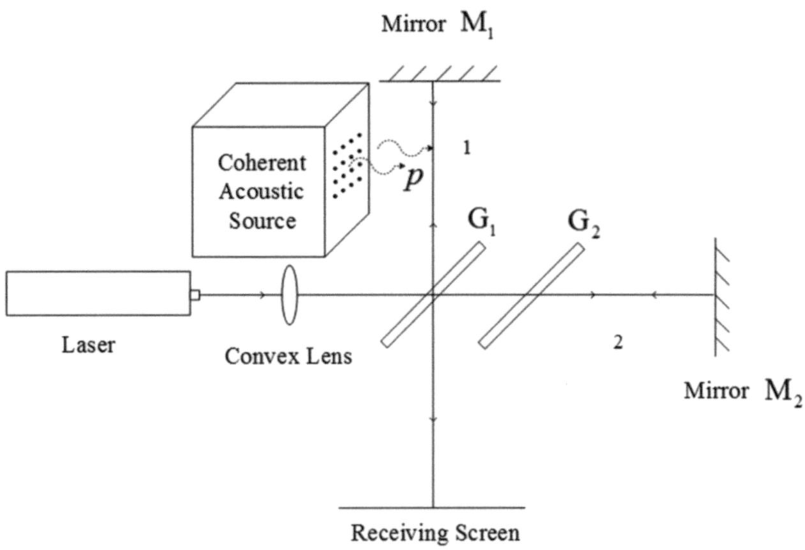

Figure 21. Artificially changing the phase of the electromagnetic wave with the help of the coherent acoustic wave-induced atmospheric refractivity fluctuation.

through the normal air. Both beams meet at a receiving screen to produce an interference pattern. Comparing the interference fringes when the acoustic source is on and off, the feasibility of artificially changing the phase of the electromagnetic wave is demonstrated.

As shown in **Figure 9**, the 7 × 7 hole acoustic excitation system working at 300 Hz is adopted in the testing experiment; its parameters are shown in **Table 1**. Before the acoustic source is on, a dark fringe is in the center of the interference fringes as shown in **Figure 22**. After the acoustic source is on, the variation of the interference fringes is captured by a high-resolution vidicon. The typical frame pictures are extracted by a video processing software, which are shown in **Figures 23–26**. **Figures 23–26** obviously show that the central fringe is gradually changing. So the experimental results qualitatively verify the feasibility to purposefully change the phase of the electromagnetic wave by the artificial refractivity fluctuation.

Parameter	Total power of the horn speaker	Diameter of the horn speaker	Diameter of the inner cavity hole	Diameter of the external cavity holes
Value	5 W	0.10 m	5 mm	2 mm
Parameter	Structure	Distance between the front panels of the inner and external cavities		Interval of the external cavity holes
Value	7 × 7	1.2 m		30 mm

Table 1. The parameters of the 7 × 7 hole coherent acoustic source in **Figure 9**.

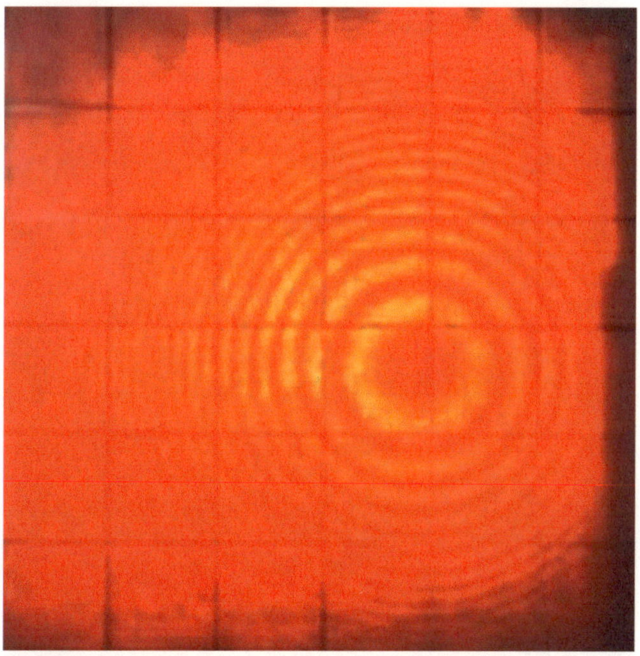

Figure 22. The fringe pattern of Michelson interferometer before the acoustic source is on.

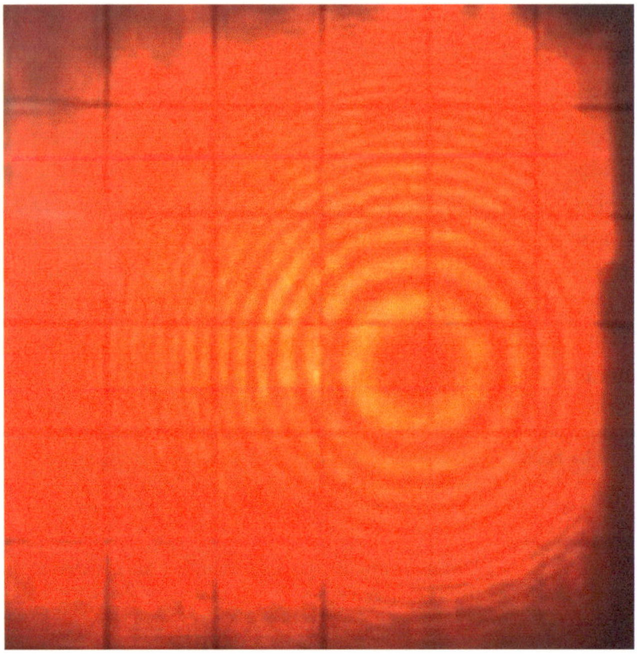

Figure 23. The fringe pattern 1 of Michelson interferometer after the acoustic source is on.

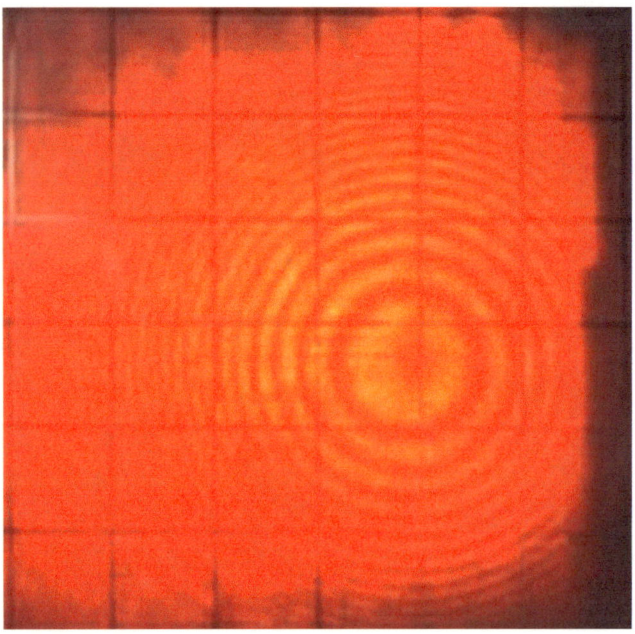

Figure 24. The fringe pattern 2 of Michelson interferometer after the acoustic source is on.

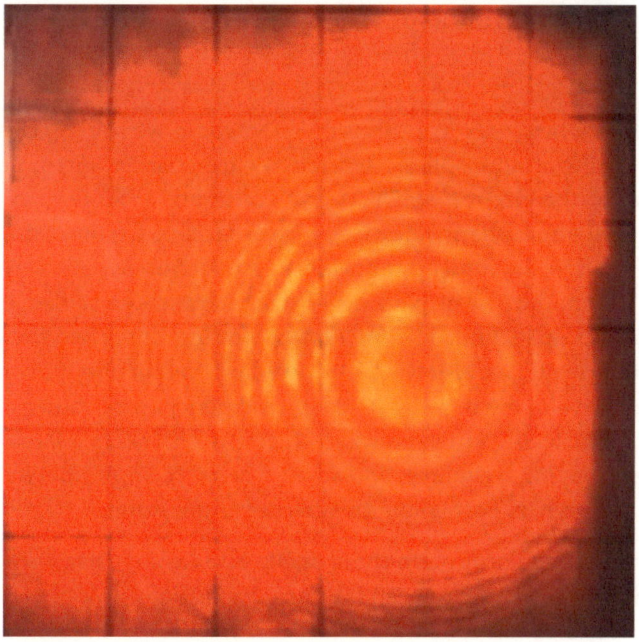

Figure 25. The fringe pattern 3 of Michelson interferometer after the acoustic source is on.

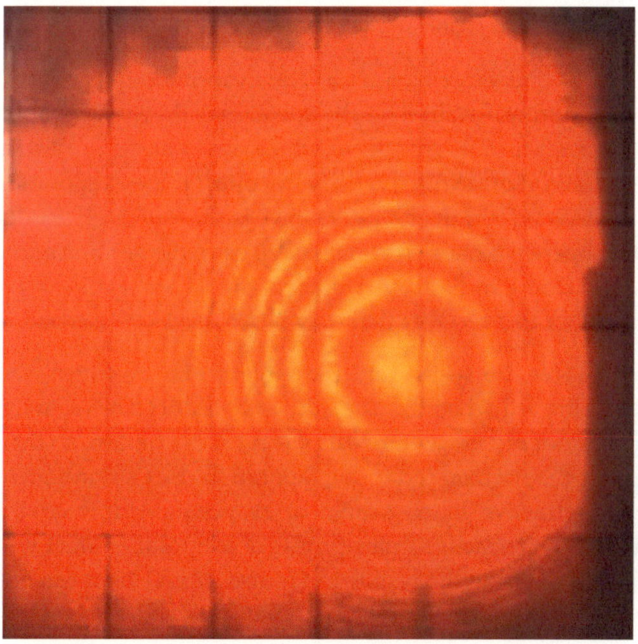

Figure 26. The fringe pattern 4 of Michelson interferometer after the acoustic source is on.

4. The potential application aspects of the artificial refractivity fluctuation

This chapter proposes a new way to purposely affect radio wave propagation by controlling and adjusting the artificial atmospheric refractivity irregularity distribution. Based on the accurate calculation of the electromagnetic wave scattering, reflection, and refraction in the areas with the artificial atmospheric refractivity irregularities, the amplitude, phase, propagation direction, polarization, and other parameters of an electromagnetic wave can be controlled artificially, so that we can purposely exert the environment media-caused positive or negative effects on radio wave propagation. The method of altering the properties of electromagnetic waves by a coherent acoustic wave-induced tropospheric refractivity fluctuation is possible to realize leap-forward development in the fields of emergency troposcatter communication, the over-the-horizon radar stealth in evaporation duct environments, electromagnetic countermeasure of coherent imaging technique, and so on. The more applications need to be explored and discovered by researchers harnessing their clever and wisdom.

The artificial dielectric irregularities are more stable and lasting than the atmospheric turbulence, and its distribution construction can be steered artificially. It is obvious that the artificial dielectric irregularities can be used to effectively improve the stability of the troposcatter communication channel. Based on the experiment shown in **Figures 23**, **24, 25**, and **26**, if the amount of phase shift caused by the coherent acoustic waves is quantitatively calculated, the interference fringes on the receiving screen can be artificially controlled. In a similar way, the artificial dielectric irregularities can be used to disturb the coherent optical imaging radar. In what follows, the mechanism of the application to the over-the-horizon radar stealth in evaporation duct environments is briefly discussed.

Because of some natural phenomena, such as the evaporation of sea water, heat exchange between air, and sea surface, as the height increases, the water vapor pressure rapidly decreases, and the temperature increases, which causes a steep fall in N. If the refractivity gradient meets the condition of $\mathrm{d}N/\mathrm{d}h < -0.157$ N - unit/m, the evaporation duct will arise in the specific region. If the wavelength of a radio wave is short enough and its angle of elevation is small enough, the super-refraction of radar wave over the sea surface will occur. In other words, the evaporation duct can be regarded as a natural waveguide which steers the radio signal from the transmitter to a receiver that may be situated well beyond the radio horizon.

If coherent acoustic sources are set around the target to be protected, the evaporation duct environment will be destroyed when the acoustic wave-induced refractivity fluctuation makes the refractivity gradient not accord with $\mathrm{d}N/\mathrm{d}h < -0.157$ N - unit/m, so that the target will not be detected by the over-the-horizon radar in evaporation duct environment. **Figure 27** shows the simulation of the refractive index profile of evaporation duct environment. **Figure 28** shows the simulation of the refractive index profile of evaporation duct when coherent acoustic waves are applied. Comparing the results in **Figures 27** and **28**, it can be concluded that the refractive index profile of evaporation duct is easily destroyed by coherent acoustic waves.

The potential applications proposed in this section based on the mechanism altering the properties of an electromagnetic wave by coherent acoustic wave-induced tropospheric refractivity

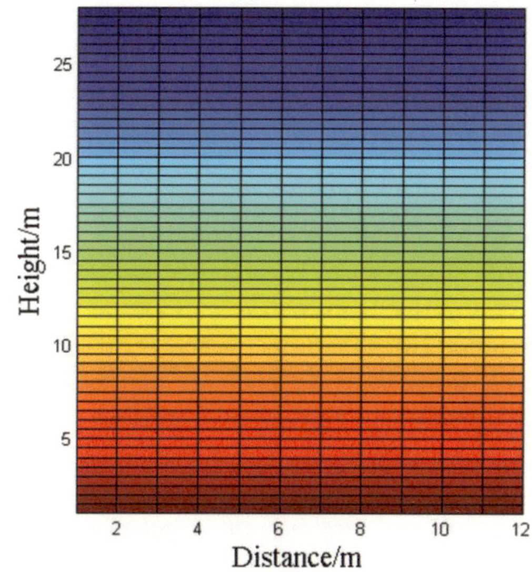

Figure 27. The refractive index profile of evaporation duct environment.

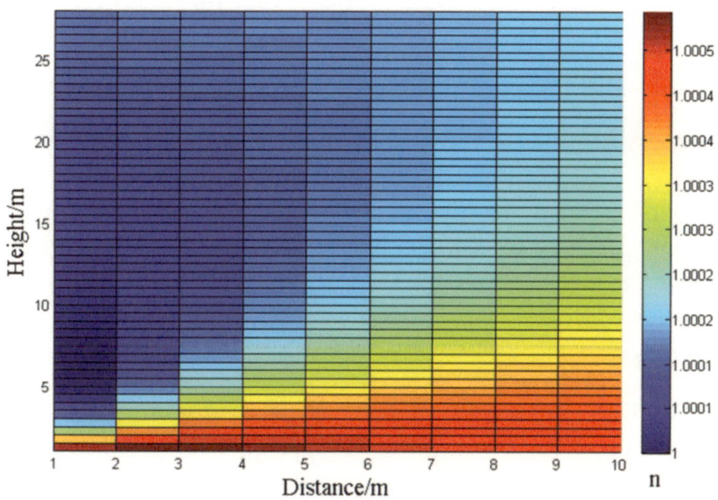

Figure 28. The refractive index profile of the disturbed evaporation duct environment.

fluctuation are still in the infant stage. There is still a long way to go for the technology development and more applications.

5. Conclusion

The idea of this chapter is logical in theory, and its feasibility is qualitatively verified in [18]. The mechanism the artificial dielectric irregularities impacting radio wave propagation is similar to that the atmospheric turbulence impacting radio wave propagation. The Bragg volume scattering theory and Fresnel volume scattering theory provide the theoretical basis for quantitatively analyzing the influences of the artificial irregularities on the amplitude, phase, propagation direction, polarization, and other parameters of an electromagnetic wave. The experiment that effectively controlled Michelson interference fringes by the coherent acoustic waves further qualitatively verifies the feasibility of purposefully changing the phase of the electromagnetic wave by the artificial dielectric irregularities. The technique of altering the properties of an electromagnetic wave by coherent acoustic wave-induced tropospheric refractivity fluctuation is very possible to realize leap-forward development in the fields of emergency troposcatter communication and the over-the-horizon radar stealth in evaporation duct environments. However, the investigation is in its infant stage; there is still a long way to go for the technology development and more applications. In order to further investigate the point of view put forward in this chapter, the following issues need to be considered deeply: quantitatively investigating the spatial distribution of atmospheric refractivity under the interaction of coherent acoustic waves and atmospheric turbulence, determining the specific influence mode of the artificial dielectric irregularities on radio wave propagation, such as reflection, scattering, refraction, and diffraction, quantitatively calculating the physical and electromagnetic properties of the artificial dielectric irregularities, quantitatively estimating the impact degree of the artificial dielectric irregularities on radio wave propagation, and discovering more application modes of this technique.

Acknowledgements

The authors would like to thank the support of National Natural Science Foundation of China (grant no. 61771375), and they would also like to thank the teachers and students who are devoted to exploring the idea of this chapter.

Author details

Shuhong Gong[1,2]*, Yu Liu[1,2], Muyu Hou[1,2] and Lixin Guo[1,2]

*Address all correspondence to: ljbrp2003@aliyun.com

1 School of Physics and Optoelectronic Engineering, Xidian University, Xi'an, China

2 Collaborative Innovation Center of Information Sensing and Understanding, Xidian University, Xi'an, China

References

[1] Tonning A. Scattering of electromagnetic wave by an acoustic disturbance in the atmosphere. Applied Science Res. 1957;**6**(1):401-421. DOI: 10.1007/BF02920397

[2] Bhatnagar N, Peterson A. Interaction of electromagnetic and acoustic waves in a stochastic atmosphere. IEEE Transactions on Antennas and Propagation. 1979;**AP-27**:385-393. DOI: 10.1109/TAP.1979.1142091

[3] Frankel M, Chang N, Sanders M Jr. A high-frequency radio acoustic sounder for remote measurement of atmospheric winds and temperature. Bulletin American Meterological Society. 1977;**58**:928-933. DOI: 10.1175/1520-0477(1977)058<0928:AHFRAS>2.0.CO;2

[4] Marshall J, Peterson A, Barnes A Jr. Combined radar-acoustic sounding system. Applied Optics. 1972;**II**:108-112. DOI: 10.1364/AO.11.000108

[5] Max North E, Peterson AM, Dean Perry H. RASS, a remote sensing system for measuring low-level temperature profiles. Bulletin of the American Meteorological Society. 1973;**54**(9): 912-919. DOI: 10.1175/1520-0477(1973)054<0912:RARSSF>2.0.CO;2

[6] Daas M, Knochel R. Microwave-acoustic measurement system for remote temperature profiling in closed environments. In: Microwave Conference; 5-9 Sept. 1992; Helsinki, Finland. IEEE; 1992. p. 1225-1230. DOI: 10.1109/EUMA.1992.335871

[7] WeiB M, Knochel R. A Monostatic radio-acoustic sounding system. In: Microwave Symposium Digest, 1999 IEEE MTT-S International; 13-19 June 1999; Anaheim, CA, USA. IEEE. p. 1871-1874. DOI: 10.1109/MWSYM.1999.780338

[8] WeiB M, Knochel R. A Monostatic radio-acoustic sounding system used as an indoor remote temperature profiler. IEEE Trans. on Instrumentation and Measurement. 2001;**50**(5): 1043-1047. DOI: 10.1109/19.963155

[9] Brooker G, Martinez J. Low-cost Monostatic radio-acoustic sounding system for indoor temperature profiling. In: Radar Conference; 26-30 May 2008; Rome, Italy. IEEE; 2008. p. 245-250. DOI: 10.1109/RADAR.2008.4720764

[10] Saffold J, Williamson F,Ahuja K,Stein L, Muller M. Radar-acoustic interaction for IFF applications. In: Radar Conference; 22-22 April 1999; Waltham, MA, USA. USA: IEEE; 1999. p. 198-202. DOI: 10.1109/NRC.1999.767313

[11] Hanson J, Marcotte F. Aircraft wake vortex generation using continuous-wave radar. Technical Digest. 1997;**18**:348-357

[12] Marshall RE, Myers TJ. Wingtip generated wake vortices as radar targets. In: Radar Conference; 13-16 May 1996; Ann Arbor, MI, USA. IEEE; 1996. p. 184-189. DOI: 10.1109/NRC.1996.510678

[13] Alexander SP. Studies of the Lower Troposphere [thesis]. Adelaide, Australia: University of Adelaide; 2004. p. 34

[14] Peters G. History of RASS and its use for turbulence measurements. In: Geoscience and Remote Sensing Symposium, 2000. Proceedings. IGARSS 2000. IEEE 2000 International; 24-28 July 2000; Honolulu, HI, USA. USA: IEEE; 2000. p. 1183-1185. DOI: 10.1109/IGARSS.2000.858061

[15] Angevine WM. Radio acoustic sounding system (RASS) applications and limitations. In: Geoscience and Remote Sensing Symposium; 24-28 July 2000; Honolulu, HI, USA. USA: IEEE; 2000. p. 1180-1182. DOI: 10.1109/IGARSS.2000.858060

[16] Ulyanov YN, Maksimova NG, Misaylov VL. Radioacoustic sounding with the use of wideband acoustic impulses. In: Ultrawideband and Ultrashort Impulse Signals; 15-19 Sept. 2008; Sevastopol, Crimea, Ukraine: IEEE; 2008. p. 249-251. DOI: 10.1109/UWBUS. 2008.4669425

[17] Furumoto J-i, Shinoda T, Tsuda T. Continuous monitoring of temperature profiles with a excellent vertical resolution by applying frequency domain Interferometric imaging technique to the radio acoustic sounding system with a wind profiling radar. In: ICROS-SICE International Joint Conference; 18-21 Aug. 2009; Fukuoka, Japan: IEEE; 2009. p. 1312-1315

[18] Gong S, Yan D, Wang X. A novel idea of purposefully affecting radio wave propagation by coherent acoustic Source-induced atmospheric refractivity Fluctuation. Radio Science. 2015;**50**(10):983-996. DOI: 10.1002/2015RS005660

[19] Hall MPM. Atmospheric Refractivity, in Tropospheric Propagation and Radio Communication. translated by Z. Liang, and Z. Zhang. Beijing:1984

[20] Gong S. Study on some Problems for Radio Wave Propagating and Scattering through Troposphere [Thesis]. Xi'an, China: Xidian University; 2008

[21] Du G, Zhu Z, Gong X. Basic Properties of Acoustics, in Basic of Acoustics. Shanghai, China: Shanghai Sci. and Technol; 1991

[22] Ishimaru A. Wave Propagation and Scattering in Random Medium. NewYork, NY, USA: Academic press; 1978

[23] Kallistratova MA. Backscattering and reflection of acoustic waves in the stable atmospheric boundary layer. In: 14th International Symposium for the Advancement of Boundary Layer Remote Sensing; IOP Conf.Series:Earth and Environmental Science; 2008. DOI: 10.1088/1755-1307/1/1/012001

[24] Kirkwood S, Belova E, Satheesan K, Narayana Rao T, Rajendra Prasad T, Satheesh Kumar S. Fresnel scatter revisited – Comparison of 50MHz radar and Radiosondes in the Arctic. The Tropics and Antarctica. 2010;**28**:1993-2005. DOI: 10.5194/angeo-28-1993-2010

[25] Jagdhuber T. An approach to extended Fresnel scattering for modeling of depolarizing soil-trunk double-bounce scattering. Remote Sensing. 2016;**8**:818. DOI: 10.3390/rs8100818

[26] Gage KS, Balsley BB, Green JL. Fresnel scattering model for specular echoes by VHF radar. Radio Science. 1981;**16**(6):1447-1453. DOI: 10.1029/RS016i006p01447